CITIZEN SCIENCE
GUIDE FOR FAMILIES

Huron Street Press proceeds support the American Library Association in its mission to provide leadership for the development, promotion, and improvement of library and information services and the profession of librarianship in order to enhance learning and ensure access to information for all.

CITIZEN SCIENCE

GUIDE FOR FAMILIES

Taking part in real science

GREG LANDGRAF

an imprint of the American Library Association

HURON STREET PRESS

CHICAGO 2013

GREG LANDGRAF is a Chicago-based writer and a volunteer at the Peggy Notebaert Nature Museum. See his website at http://greglandgraf.wordpress.com.

Printed in the United States of America

17 16 15 14 13   5 4 3 2 1

Extensive effort has gone into ensuring the reliability of the information in this book; however, the publisher makes no warranty, express or implied, with respect to the material contained herein.

ISBNs: 978-1-937589-35-6 (paper); 978-1-937589-36-3 (PDF); 978-1-937589-37-0 (ePub); 978-1-937589-38-7 (Kindle). For more information on digital formats, visit the ALA Store at alastore.ala.org and select eEditions.

Library of Congress Cataloging-in-Publication Data

Landgraf, Greg.
    Citizen science guide for families : taking part in real science / Greg
    Landgraf.
        pages   cm
    Includes bibliographical references and index.
    ISBN 978-1-937589-35-6 (pbk.)
        1. Research—United States—Citizen participation. I. Title.
Q180.55.C54L36 2013
500—dc23
                                                    2013010982

Book design by Kimberly Thornton in Minon Pro, Popular, and Gotham.
Cover photograph © Phaitoon Sutunyawatchai/Shutterstock.
♾ This paper meets the requirements of ANSI/NISO Z39.48–1992 (Permanence of Paper).

For Jeff, in hopes that RHIC will eventually be able to accelerate a grapefruit

*With special thanks to Jill Doub, Val Hawkins, Leonard Kniffel,
Fred and Sandy Landgraf, Laura Saletta, the staff and volunteers of the
Peggy Notebaert Nature Museum, and all the citizen scientists and citizen science
project coordinators who have shared their stories for this book*

CONTENTS

FOREWORD

CONTEMPORARY CITIZEN SCIENTISTS CITE THE CHRISTMAS BIRD COUNT OF 1900 as the beginning of the general public observing and recording scientific data. But even before that moment, people around the world were making formalized observations of the nature around them. Benjamin Franklin's archives contain famous examples of phenological records—notes about when local plants were flowering—but farmers and city dwellers alike frequently recorded such observations as well. These records, created by thoughtful and observant people from all walks of life, are irreplaceable data that help us all understand modern issues such as the spread of some species and the loss of others, from the impacts of local habitat fragmentation to the causes of global climate change.

However, even before the systematic collection of data for scientific use, our ancestors were observing moon phases, noting the extent of their local watershed, and marking the migration time of animals. These are fundamental human activities that connect us to the natural world, and indeed to one another. In an ever more urbanized and disconnected society—one that increasingly relies on technology for social connection—our need for a direct connection with the natural world is critical. Participating in citizen science programs brings people out into that natural world, whether it is their own backyard, an urban park, or a remote nature preserve.

As the "urban gateway to nature and science," the goal of the Chicago Academy of Sciences and its Peggy Notebaert Nature Museum is to connect people to the natural world around them. Our citizen science projects not only involve some of the most interesting life forms on our planet, but they provide an impetus for people in this urban environment to recognize and appreciate the nature they live with. Citizens' active participation provides critical data that inform meaningful scientific progression and conservation activities.

Greg Landgraf is one of those active participants in our shared natural world. His passion and enthusiasm has been a bridge between museum guests to their environment for the last several years. He is an extraordinary volunteer and champion of citizen science. He has a special talent of conveying the excitement one gets out of participating in our citizen science programs while working with museum guests.

Citizen science is not a new phenomenon; it is however, revitalized by people yearning to reconnect with nature. Today both scientists and citizens alike are using this process in a more rigorous and coordinated manner that will elicit new knowledge that contributes to scientific progression and to a better place for all of us to live.

Deborah Lahey
President & CEO
Peggy Notebaert Nature Museum,
   Chicago Academy of Sciences

Steve Sullivan
Senior Curator of Urban Ecology
Peggy Notebaert Nature Museum,
   Chicago Academy of Sciences

PREFACE

LINCOLN PARK'S NORTH POND IS ONE OF MY FAVORITE PLACES IN CHICAGO.
It's part of a wonderfully picturesque park, with plenty of trees, open space for recreation, and a few statues with carefully planted flower beds. But the pond itself is even better: a place to experience nature, even in the middle of the city. Most of my days begin with a stroll around the pond, which I find to be an ideal way to clear my mind for whatever rigors the day ahead may hold.

The pond and its immediate surroundings, which include significant swaths of restored native prairie, attract all sorts of wildlife—a pair of great blue herons, dozens of basking red slider turtles, butterflies, dragonflies, rabbits, and much more. But as delightful as all of these creatures are to view, some days I have a different mission.

My quarry is that oh-so-common critter, the gray squirrel. They aren't hard to find. I spot one scampering up a maple tree, a couple more chasing each other along the top of a park bench, a few more chewing on seeds that have fallen to the ground, and a jackpot of several squirrels scurrying around a copse of trees near the end of my journey. Today, I've spotted thirty-one squirrels in all.

This may seem like an unimportant detail, but there's a good reason for counting squirrels. When I get home, I visit projectsquirrel.org, the website

The author in Chicago's Lincoln Park, collecting data for Project Squirrel.

Photo by Laura Saletta

for Project Squirrel, and share the number of squirrels I've counted, as well as a few details about the site I spotted them and the date and time of my walk.

Project Squirrel is an example of citizen science, a project organized by scientists but that welcomes observations from interested members of the general public. These projects are generally organized around important and interesting scientific questions: whether frog populations are stable or dropping precipitously, what causes dragonflies to swarm, or the best way to manage natural resources to ensure the health of animal populations. Among other things, Project Squirrel seeks to determine why gray squirrels and the smaller fox squirrel typically don't live in precisely the same areas, even though they could.

It may seem surprising that ordinary citizens can contribute to this or any kind of scientific research. The stereotype of a scientist, after all, involves frizzy hair, speech consisting of nothing but incomprehensible gobbledygook, and sterile laboratories full of complicated and expensive equipment, doing work that no mere mortal could hope to comprehend without years of training and an impossible level of inborn genius.

Certainly some science does require advanced training or expensive equipment. But there's also a lot that we don't know about in our own backyards because there just aren't enough people looking to see what's there.

Take, for example, the nine-spotted ladybug. We know that it can live throughout the United States and southern Canada from ample historical records of it. But populations have declined sharply in the Northeast. It's no longer found all the way through the country, and more important, we don't know all of the places where it can be found. There simply aren't enough scientists to scour the land and catalog every possibility.

And so, even though it's the state insect of New York, the nine-spotted ladybug was believed for almost thirty years to be extinct within the state.

That changed in 2011, and it's not because of the efforts of professional researchers. It was a citizen scientist collecting insects for a citizen science effort called the Lost Ladybug Project who rediscovered the nine-spotted ladybug on a farm in Suffolk County.

There are hundreds of citizen science projects out there on dozens of topics. There's also an incredible range in terms of the time commitment they require (from just a few minutes to regularly recurring surveys that last several hours) and the level of skills and knowledge you need to participate. Most include educational components to help participants learn about their subject, as well as advocacy efforts to help build support for the conservation of rare or endangered plants or animals, although the mix of research, education, and advocacy varies widely as well.

Many projects, however, require only the most basic skills and the minimal time it takes to make an observation. And most are convenient, with many seeking information about what you can see in your backyard or neighborhood.

Because of that, citizen science is a wonderful thing to share with your family. Even young children are welcome to participate in many projects. In fact, their energy and sharp eyesight can be big advantages when hunting elusive dragonflies or salamanders. One of the Lost Ladybug Project's first major discoveries was made by a brother and sister, ages 10 and 11, finding a nine-spotted ladybug near their home in Virginia in 2006—the first of the species seen in the eastern United States in fourteen years.

This book will examine in depth the reasons for taking part in citizen science, profile several active participants in citizen science programs and share their reasons for volunteering their time, and provide a guide to help you find the right citizen science project for you and your family.

Before getting to that, though, I'd like to briefly extol the virtues of the trait that I think is at the heart of all science: curiosity.

My favorite expression of this trait in all of human culture comes from the song "Still Alive," from the video game Portal. It plays after you've defeated the big boss, an evil supercomputer named GLaDOS that has been "experimenting" on you, and won the game. But even though it seems like she has been incinerated, well, she's not *quite* dead. "I'm doing *science,* and I'm *still alive,*" she taunts.

While obviously evil is . . . well, evil, and not something to aspire to, I can't help but admire the notion that doing science is the same as being alive. The opportunity to explore, and to learn, and to remain constantly curious is one of the greatest parts of being alive, and should be relished.

There's a lot to discover in this world, and citizen scientists have every right to discover it. Enjoy the search!

CHAPTER 1

Being a Citizen Scientist

VERYONE KNOWS THE STEREOTYPE OF THE SCIEN- tist from the movies: the impossibly brilliant genius with frizzy hair and white coat who spends his time locked in a lab with twisty bits of glassware and a Jacob's Ladder generating high-voltage electric sparks at regular intervals.

Reality is, of course, quite different. Science and scientific research are far more accessible than the stereotype might imply. In fact, professional science itself is a relatively new concept. Despite their incredible contributions to science, Benjamin Franklin, Charles Darwin, and Henry Walter Bates could all be considered amateur scientists, planning and funding their own investigations rather than conducting them on behalf of an organization.

The process of science is truly open to anyone. "Science is thought of as something that other people do, but really, it's a set of steps to follow to answer a question," says Karen Purcell, a scientist at Cornell University. "You don't need a PhD or a masters for that."

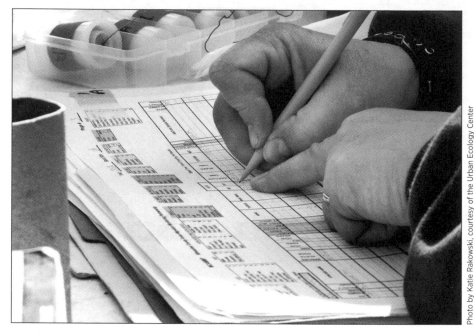

Careful data recording is critical to all scientific endeavors.

A lot of today's scientific research does require sophisticated equipment and advanced, specialized knowledge—but not all of it does. In many cases, what researchers need to find answers to their questions is more observations from more places than they could go themselves. Purcell, for example, is coordinator of a project called Celebrate Urban Birds that collects observations from anyone who wants to contribute about how birds use green space in urban areas.

Celebrate Urban Birds is an example of citizen science. Also sometimes called "public participation in scientific research," citizen science can take a number of forms, including independent research projects, research expeditions that welcome interested amateurs, or science hack days that bring people together for brief, intense periods of collaboration.

This book focuses on citizen science that is accessible to all ages and all levels of scientific knowledge. This encompasses a large number of citizen science projects. These projects are organized by some individual or group, usually a professional researcher, but designed with protocols that amateurs can follow to either collect or analyze data to contribute to the research.

This kind of project has a long history. The first example is often considered the Audubon Society's Christmas Bird Count, which started in 1900 when ornithologist Frank Chapman proposed ending the tradition of Christmas side hunts and counting birds instead. And while there may be fewer people willing and able to fund their own scientific expeditions now, the number of these citizen science projects has grown dramatically in recent years.

The Internet is a major contributor to this growth. Modern websites make it fast and easy for project participants to report data. Many projects also release data through their websites for anyone, amateur or professional, to view. This data can often be displayed in highly customizable ways so researchers can focus on an individual species or region of interest.

Mobile technology is also helping spur growth. While not yet universal, many citizen science projects have created apps so participants can document their observations with a photograph and submit them to the project through their smartphones.

One of the great benefits of citizen science is that it gives its participants a glimpse of the scientific process. While science has the reputation of incredible complexity that requires years of training to be able to take part in, the basic components of that process—making observations, recording them, and contributing them to a larger set of data—are accessible to anyone.

Because of that accessibility, anyone can be valuable to science. The world is big, and there is plenty in nature that hasn't been observed yet. In many cases, the reason isn't that things are hard to see, but rather that nobody has been able to look yet.

"There's no way in the world we could replicate with two or three grad students what this workforce of citizens is contributing," says Karen Wilson, who founded the Michigan Butterfly Monitoring Network and now works with the Illinois Butterfly Monitoring Network. "And the more information they gather, the better for making decisions."

"For a project that needs a lot of data, having a citizen science component makes a lot of sense," says Andrea Lucky, head of the School of Ants project where citizen scientists collect samples of ants for identification and species mapping. And most scientific research does require a lot of data. The epiphany that led to the theory of gravity may have come to Isaac Newton when he saw an apple falling from a tree, but it took him two decades to fully develop the theory. (And that theory was superseded by Einstein's theory of relativity, and research continues to this day to explain phenomena that Einstein's theory doesn't.)

"It's many small observations over time that lead to significant truths. It's not an aha moment every day," says Jeremy Gieser, director of tours and marketing for Gastineau Guiding, an Alaska company that offers tours with a citizen science component.

And more citizen scientists means more science. "The more folks that are out there, engaged and paying attention and letting people know what they see, the more we can learn about our world," adds David Bonter, leader of Project FeederWatch, which gathers long-term data about birds that use feeders in the winter.

Crowdsourcing some aspects of research gives scientists the benefit of a large workforce to collect and analyze data in ways that individual researchers can't. The scope of that data is necessary to understand events that occur on a global scale such as climate change. "The scale of citizen science research matches the environmental issues that we need to address," says Janis Dickinson, director of citizen science at the Cornell Lab of Ornithology, which operates several citizen science projects.

But what is the benefit to the participant? That depends on the project and especially the participant, but there are a lot of possibilities. In many cases, citizen science taps into a preexisting passion. If there's an animal, plant, habitat, or phenomenon that you're interested in, there is likely a citizen science project that will provide a good excuse to study it regularly—as well as direction to study it in ways you might not have considered. For example, Mike Newkirk began volunteering with the Vermont Loon Recovery Project after realizing how much he missed the majestic loon calls that he heard at camp in the Adirondacks as a child.

In many cases, the gap between hobby and science is slim. Many citizen science projects, for example, aim to determine which species of animals live where. Bird-watchers, butterfly watchers, and other nature enthusiasts already have the observation and identification skills necessary to collect this kind of data, so participation in science doesn't require much more than sharing it.

Even nonexpert amateurs may already be making valuable observations. Ever record the ice-out date of a nearby pond or the day the first robin returns in spring? Those are important observations of phenology—the study of periodic life-cycle events—that many researchers and several citizen science projects are using to study the effects of climate change. "A lot of people don't even realize that they're doing citizen science when they do that," says Michele Tremblay of the Upper Merrimack Monitoring Program.

Contributing to citizen science is a valuable educational opportunity for young and old alike. Participants have the opportunity to learn about the subject they are studying, but also about nature more generally and about the scientific process. "We live in a scientifically illiterate society at the moment," says Christine Goforth founder of the Dragonfly Swarm Project. "Getting people while they're young is valuable. They'll be more inclined to learn about science in the future and become scientists themselves."

That scientific understanding is vital for the public to be able to contribute to decision-making about natural resources in their communities. In fact, while most citizen science projects were created by professional scientists, there are several examples of projects led by nonscientists or by scientists working outside of their specialty with the primary goal of studying and improving natural areas in their communities. Among them are the Southwest Monarch Study in Arizona, the PRIDE Shorebird Survey on the Lower Santa Ynez River in California, and the Clark County Amphibian Monitoring Project in Washington. See chapter 5, "Citizen Science Making a Difference," for more about community-focused citizen science projects.

One final factor shouldn't be overlooked: being outside and exploring nature is a pleasurable experience. "It is fun to find monarchs in the wild. It inspires young and old to go outside and explore nature, even if it is just their backyard," says Wendy Caldwell, community program assistant for the Monarch Larva Monitoring Project. "Once they start monitoring for monarchs, most volunteers become hooked and spread the word to their friends, family, and neighbors, who are inspired to do the same."

Most citizen science projects balance three broad goals: research, public education, and advocacy for the protection and conservation of the subjects they study. The specific balance of these three factors can vary widely. Some projects are as concerned or more concerned with their educational goals as research, and use the act of data collection as a teaching mechanism; others sacrifice the accessibility that might help them attract and educate a wider participant base for rigid scientific protocols.

Can citizen science actually generate reliable, scientifically useful data? The answer is yes. Hundreds of peer-reviewed scientific papers have cited citizen science data, according to a July 31, 2012, National Science Foundation webcast (www.nsf.gov/news/news_videos.jsp?cntn_id=124991&media_id=72892&org=NSF).

The question isn't quite as simple as that, though. Naturally, amateurs can make mistakes in collecting data, by misidentifying what they see or following the wrong procedures to collect information. Not surprisingly, data quality is a big concern for many researchers who have organized citizen science projects. "If citizen science can't produce high-quality data, scientists won't be interested in it for much longer," says John Pickering, founder of Discover Life, which operates the Mothing citizen science project.

But researchers have come up with several ways to ensure the information their projects collect is valid.

Most projects offer training of some sort. In its most basic form, generally for relatively simple projects that only require observations of a few, easy-to-identify phenomena, this might take the form of some basic online materials for participants to review. More in-depth projects offer online courses or webinars, or in-person training sessions both in a classroom setting and in the field. Some projects, particularly surveys that require identification of birds or amphibians by sound, provide materials that need to be studied independently. Some projects, including those that study animals that might be hurt by improper human contact, require participants to pass a test or achieve certification before they are allowed to proceed.

The Coastal Observation and Seabird Survey Team project, in which participants identify beached birds on Pacific beaches from California to Alaska, requires a six-hour training session where volunteers learn to identify bird species and the protocols for collecting data about them. Thanks in part to these training sessions, COASST Executive Director Julia Parrish has confidence in the accuracy of volunteer-collected data. "[Volunteers are] out collecting data in a standardized fashion," she says. "That means those data are immensely useful to science."

COASST has backed that assertion by testing samples sent by volunteers to see if their identifications were correct. These tests revealed an accuracy rate of about 86 percent among volunteers, enough to support Parrish's confidence in their value.

"There's always people who get really riled up about citizens taking scientific data," notes Michele Tremblay, director of the Upper Merrimack Monitoring Program in New Hampshire. "We believe well-trained, motivated volunteers can produce good data." The project sends random samples to a professional lab to check that there are no scientifically significant errors. Because the program's observations are used to inform potentially controversial government decisions about which bodies of water are reported as impaired under the Clean Water Act, the lab the project uses is out of state to ensure it is unbiased.

In some cases, documenting an observation may allow a person to make a valuable contribution to science even if they don't know what it is that they're looking at. Several citizen science projects, such as iNaturalist, require photos with all observations. That way, the identity of the animal seen can be confirmed by experts in the project's community. In some cases, the person who contributes a photo misidentifies it; those cases become learning opportunities as the communities help correct the error.

The site has a system in place to ensure that researchers do not rely upon observations that may not be correct. All observations are placed into one of two categories: casual or research grade. "'Research grade' status on iNaturalist requires a date, location, and community agreement on the identification," says Ken-ichi Ueda, co-director of the site. "There are very few false positives."

Documentation isn't a viable option for all citizen science projects, however. Illinois Butterfly Monitoring Network participants, for example, count and identify every butterfly they see along a set route. Several species are extremely abundant on these routes; it would be impossible to attempt to snap a photo of them all, and likely not terribly interesting either.

IBMN relies upon its participants' training and experience to ensure that the information they collect is accurate. "We've got people who have been observing for ten to fifteen years," says Doug Taron, IBMN director. "Even in academic research, a lot of the data collection is done by undergrads" with less experience than program volunteers, he adds.

Citizen science can be used for both collecting and analyzing data, although the majority of projects use the public for data collection. There are two basic kinds of data collection methods. Some projects welcome opportunistic data, which are observations that can be made by participants whenever they see a plant or animal of interest to the project. Other projects require participants to follow a formal protocol, which often entails collecting data at specific times on specific routes, and following the same observation techniques each time.

Both types have their place. Opportunistic observations are usually easier entry points to citizen science. Many people, in fact, discover citizen science projects when they see something in nature that they can't identify and search online for clues. They don't generally require large time commitments, and many such observations can be made any time you are out enjoying nature. Some projects are even more convenient, explicitly seeking observations from your neighborhood or even your backyard.

Projects with stricter protocols have the benefit of scientific rigor. Many projects would be scientifically useless if they didn't follow these protocols. Many citizen science projects that seek to perform counts of specific species, for example, carefully specify the amount of time each participant should spend observing each session so that the data can be fairly compared. If a person sees five of one bird species in the field one day and ten another, that might suggest a population increase—but if the first figure is the result of ten minutes of observation and the second an hour, it may also simply reflect the difference in the amount of effort. These projects are also careful to ensure that participants report the times they go into the field but don't find what they're looking for, because those data points of zero are equally important.

Citizen science projects with strict protocols require a bit more preparation than opportunistic projects in order to successfully contribute, since participants need to familiarize themselves with the protocols. But that doesn't necessarily mean that they are difficult for beginners to be a part of. There is a wide range, but many projects have simple-to-follow protocols. School of Ants, for example, requires nothing more than a few basic supplies (a couple of cookies, a few index cards, and some plastic bags) and participants just need to set them down, watch for forty-five minutes, and collect the cards with samples of ants afterward to mail to the project's office.

Opportunistic observations can still be scientifically valuable in certain instances. Many projects seek to determine an animal's range, rather than the size of its population. Finding any example of an animal in a new place contributes important knowledge. The citizen science project All Taxa Biodiversity Inventory in the Great Smoky Mountains National Park in Tennessee, for example, has identified seven thousand species that had never been documented in the park before, and more than nine hundred of them are new to science entirely.

Other projects simply can't incorporate a strict protocol. Dragonfly swarms, for example, are ephemeral, hard-to-predict events. The Dragonfly Swarm Project, which is studying them, welcomes opportunistic observations because a strictly scheduled monitoring program would not be likely to see them. It gathers information about weather and terrain conditions in hopes of finding the kind of conditions that will attract these swarms.

Citizen science offers genuine value to scientific research while providing wonderful benefits to its participants. Although any individual observation may not lead directly to a revolution in our understanding of the world around us, they all contribute to unraveling the mysteries of life in an important way. As Bill Hilton Jr., principal investigator of the Operation RubyThroat hummingbird study, says, "You never know when that observation might be useful. When you record your observations, they become immortal."

CHAPTER 2

Science as a Family Activity

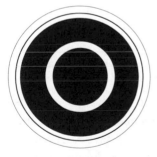

NE OF THE BEAUTIFUL THINGS ABOUT CITIZEN SCI-
ence is how it offers any individual who is interested
the opportunity to take part in scientific research.
But many projects have provisions for people to
take part in groups rather than just on their own,
and many projects are suitable for all ages. Those
two factors combined mean that citizen science can
be a dynamite activity for families to share.

Mike Newkirk says that his family's commitment to the Vermont Loon
Monitoring Project "forced us to be a good family." The project involves regular monitoring of breeding loons and maintenance of nesting floats that
provide the habitat the loons need to build nests. Family members took the
project seriously, blocking out time to visit their monitoring site together.
Newkirk says those trips helped forge strong family connections that remain
stable even though his kids are now all adults.

Most of the time, parents are the experts, the rule makers, and the guides.
But citizen science projects often involve visiting sites few people go to, or
making close observations of a plant or animal that would normally be barely

noticed. Since citizen science activities are often new experiences for parents and kids alike, all members of the family have a chance to make real and important contributions. "Kids ask questions and see things adults don't," observes Mía Monroe, coordinator of the Western Monarch Thanksgiving Count in California. Those questions benefit the parents as well, because they give an opportunity for the whole family to learn together.

"The Monarch Larva Monitoring Project is great for families," says Wendy Caldwell, the project's community program assistant. Project participants visit sites where milkweed grows to count monarch butterfly eggs and caterpillars, an activity that's easy for parents to share with their children. "A lot of families become interested in MLMP because not only do they enjoy monarchs, but their children fall in love with science, learning, and exploring nature. It is a way for everyone to learn about biology and ecology while contributing important data to such a big project," Caldwell notes.

Citizen science can help keep kids interested in the natural world at a time when many lose that interest. "There's some age where people stop liking insects and start thinking they're gross," says Christine Goforth, creator of the Dragonfly Swarm Project.

She notes that the dragonflies that she studies may have some advantage when reaching out to girls because the dragonfly is currently a common motif in fashion and jewelry, so "there's a population of girls growing up realizing that dragonflies are cool." The opportunity to study them and contribute to research about them, however, can help girls and any kids to realize that they are cool to learn about as well as to look at.

Of course, it's absurd to say that a lack of interest in the natural world is solely a problem among girls. "A lot of people get creeped out by bugs and don't have an appreciation of what they need in a habitat," says Kurt Mead, leader of the Minnesota Odonata Survey Project that studies the state's dragonflies. It's particularly important to build interest among children, however, because that interest is then more likely to carry through to adulthood.

It's critical to reach all children early, says Ron Zwerin, communications manager for the GLOBE Program, a network that connects students and teachers with scientists. "From a STEM [science, technology, engineering, and math] perspective, if we don't get to these students by fourth grade, we're going to lose them."

Taking part in citizen science affords an opportunity to see the wonders of nature up close. The educational value goes beyond learning rote facts and figures. "In nature, kids see, hear, smell and touch things all at the same time, getting them to observe, ask questions and figure out things that have a lot of parts to them," asserts the US Fish and Wildlife Service's Let's Go Outside! website (www.fws.gov/letsgooutside). Examining plants or animals in detail can help kids to learn that most things get more fascinating the more you learn about them.

Many of the subjects of citizen science projects are already well-established as part of school curriculums. "The monarch is such a well-known insect that it is used by many teachers and youth leaders in their classrooms to teach biology and metamorphosis, and inspires youth to come up with experimental questions and carry out their own explorations to find answers to these questions," Caldwell says.

There's absolutely no reason that this type of investigation needs to be limited to a classroom setting, however, and citizen science is one framework that can help parents to support their children's scientific curiosity. "I think inquiry is the best way to get kids curious and engaging more deeply in natural history observations," says Janis Dickinson, director of citizen science at the Cornell Lab of Ornithology, which operates several citizen science projects.

The effects can last in significant ways. "We've had kids enter Cornell who have been doing Project FeederWatch [one of the Cornell projects] for ten years," Dickinson says. "They become star students" and often publish their own research while still undergraduate students, she says.

Mary Jane Schramm, media and public outreach specialist for the Gulf of the Farallones National Marine Sanctuary and its Beach Watch monitoring program, says that one participant who started with the program as a child monitoring with his mother dropped out for college, only to rejoin after graduation. Laura Molenaar, an elementary school teacher who has conducted summer sessions with her students to gather data for the Monarch Larva Monitoring Project for more than a decade, says that many former students are now studying the sciences in college.

Learning isn't and shouldn't be limited to the young. "Every person enjoys seeing something new or understanding something they've seen but didn't know much about," says Todd Witcher, executive director of Discover Life in America, which coordinates the All Taxa Biodiversity Inventory that documents plants and animals in the Great Smoky Mountains National Park.

The family orientation of many citizen science projects isn't limited to just parents and their kids. Many projects report grandparents sharing monitoring duties with their grandchildren, including the Water Action Volunteers'

Citizen Stream Monitoring program in Wisconsin and the NatureWatch suite of projects in Canada. "We often get grandparents who want to share nature with their grandkids," says NatureWatch manager Marlene Doyle.

The simple act of sharing nature is one important benefit that many citizen science projects offer. "There is a concern that kids are not spending enough time outside," says Andrea Lucky, head of the School of Ants project that maps ant species across the country. "Kids won't know anything about wild-life unless they're exposed to it, especially for kids in cities." Taking a few min-utes—or an hour—or blocking out a regular time to get outside can go a long way toward helping kids to appreciate the joys that being outside can provide.

But the benefits go beyond simple pleasure. "Intellectual and physical heal-ing has been tied to green space," observes Steve Sullivan, coordinator of Proj-ect Squirrel.

According to letsmove.gov, the website for First Lady Michelle Obama's Let's Move! anti–childhood obesity campaign, "Regular exercise in nature is proven to improve children's physical and mental health. Outdoor activity helps kids maintain a healthy weight, boosts their immunity and bone health and lowers stress." The Let's Go Outside! website www.fws.gov/letsgooutside/families.html) adds: "Studies indicate that children who play and explore out-doors are less stressed and may further benefit by learning confidence and social skills. . . . Keeping kids active helps keep kids healthy. And nothing keeps kids active more than giving them fun and interesting things to explore and do."

An October 17, 2008, article in the *New York Times* reported on a study conducted at the University of Illinois at Urbana-Champaign that suggested that children with attention deficit hyperactivity disorder were better able to focus after spending time in a park than in downtown areas. "The researchers found that a 'dose of nature' worked as well or better than a dose of medica-tion on the child's ability to concentrate," writes Tara Parker-Pope in "A 'Dose of Nature' for Attention Problems." The study corroborated other ones sug-gesting that time in nature can improve psychological health.

There are also emotional benefits to being outside and communing with nature. In fact, many citizen science projects make it hard not to relax. "I don't know anyone that leaves tagging events stressed," says Gail Morris, coordina-tor of the Southwest Monarch Study that tags monarch butterflies in Arizona. She adds that participating affords an opportunity for everyone, particularly kids, to learn coping mechanisms for difficult situations. "Children can learn that when they're stressed, they can take a walk to release their anger," she says. "We have to force ourselves to do that. If we can give that to kids when they're younger, I think it will stay with them."

"For kids who grow up in this technological age, spending time in nature is a great mediator," says Marlo Perdicas, park biologist for Metro Parks, Serving Summit County in Ohio, which operates the Summit County Citizen Science Inventory. Citizen science can even help bridge the gap between technology and nature, she adds. Many citizen science projects are designed for smartphones, where participants can make their observations and take photos through the phone's camera and then upload them on the spot through dedicated apps.

Among those are Project NOAH, which allows the public to document the wildlife and plants that they see. Cofounder Yasser Ansari says that one reviewer of the project's smartphone app described how he was using it to get his daughter interested in nature. In another instance, the app allowed a family to salvage a vacation when their son broke his leg immediately beforehand. While the injury prevented him from taking part in some of the planned activities, "he used the app to document bugs and they had a blast," Ansari says.

CHAPTER 3

Building Community through Citizen Science

HILE WE MAY THINK OF SCIENCE AS AN INDIVID-
ual activity, the reality is different. Collaboration is
an important facet of the research process, as shared
knowledge, equipment, experiences, and perspectives across institutions and disciplines.

Citizen science projects such as the ones listed in
this book are inherently collaborative processes as
well, as many people (ranging from dozens to thou-
sands) each contribute their experience and effort to a greater goal. But many
citizen science projects offer a more personal form of collaboration as well:
the opportunity to build personal connections and friendships with other
participants. These types of relationships can have positive effects on physical
and emotional health; a 2010 analysis, published in *PLOS Medicine,* of almost
150 studies found that social connections can help us survive health problems, while the lack of those connections can cause them.

Most citizen science projects offer the chance to meet people who share
similar interests and passions. Those shared interests may be as valuable or
more to the participants than the scientific research they do. As Julia Parrish,

executive director of the Coastal Observation and Seabird Survey Team project that monitors beached seabirds on Pacific beaches, says: "I think what knits all COASSTers together is a strong sense of place. It has a scientific aspect but they also love the beach."

Many projects are explicitly group-oriented, designed so that participants perform their work with other people. The North American Butterfly Association, for example, organizes butterfly counts throughout the year, where local count circles may have nine or ten groups with four observers each. "It's a social thing," says NABA President Jeffrey Glassberg. "You'll get to meet other people that like nature and butterflies."

Julie Brown, monitoring site coordinator for the Hawk Migration Association of North America, says that the opportunity to monitor together is a big draw for many participants. "Hawk watchers are usually very dedicated and devoted to their favorite sites," she says. "They gather together and they share stories and laughs, all through wind and rain and snow."

The Proud Residents Investing in a Diverse Estuary Shorebird Survey monitors the Lower Santa Ynez River near Vandenberg Air Force Base, California. "The project site is fairly isolated, so it is exciting to construct a local project that can offer numerous benefits to everyone who is interested," says Tamarah Taaffe, treasurer of the La Purisima Audubon Society, which cohosts the survey with PRBO Conservation Science.

Participants in the Western Monarch Thanksgiving Count in California visit butterfly winter roosting colonies in groups of at least two. "There are people who become friends over this common interest," Project Coordinator Mía Monroe says. Many participants, she adds, will visit each other's gardens or monitoring sites as well.

Even projects where the work is normally more independent may offer the opportunity for group work. Participants in GLOBE at Night, a project that collects data about light pollution on the night sky, came together in conjunction with the National Geographic BioBlitz in Saguaro National Park near Tucson in 2011 to measure star visibility throughout the park.

Other relatively solitary projects may afford the opportunity to meet people through regularly scheduled training events. Illinois Butterfly Monitoring Network volunteers generally gather data alone, for example, but the project does host training events twice a year. "People enjoy the opportunity to get out and go into the field together," says IBMN Director Doug Taron. "On the surface it would not appear to be a social act, but the workshops take on social dimensions."

Similarly, the Minnesota Odonata (dragonfly) Survey Project hosts annual weekend workshops that offer opportunities for participants from across the state to meet one another. "By getting people together, a lot of friendships have been made between people who wouldn't have otherwise met," says project leader Kurt Mead. "They're teaching and learning from each other."

The Beach Watch monitoring program in California requires intense training, encompassing eighty hours of classroom and field work, and twenty-five people are trained at a time. "There's a real sense of family" among classmates, says Beach Watch Manager Kirsten Lindquist. Volunteers also often survey their beaches in groups.

Other citizen science projects plan or welcome group events that provide a social outlet for volunteers, even though those events may or may not collect data directly related to the project. NatureWatch, a national organization that operates a suite of citizen science projects in Canada, has many local groups that plan this type of outing. "It's validating to know that other people are interested in the same things you are," says NatureWatch Manager Marlene Doyle.

Plants of Concern is a project that monitors endangered and rare plants near Chicago. While most of the monitoring isn't suitable for large groups because of the sensitivity of many sites, the project does organize periodic forays to specific, less-sensitive sites, where participants can visit as a group to measure and count plants. "It's a real group effort and it's a lot of fun," says coordinator Susanne Masi.

The popularization of the Internet and the way it makes information sharing easier has been a big factor in the growth of citizen science. Similarly, the rise of social media has been a boon for interpersonal connections within citizen science projects. Celeste Mazzacano, project coordinator for the Migratory Dragonfly Partnership and its Dragonfly Pond Watch citizen science project, says that social media is an important way for project participants to communicate with each other. Many participants share observations of dragonfly flights on the Partnership's Facebook page, and the project is also using Twitter to help alert volunteers to migrations. "By using Twitter's instant communication to link observers, we hope to get a better idea of where these migratory swarms spend the nights, how long they fly, and whether they stop to rest and feed," Mazzacano says.

"A bit of a community has grown on our Facebook fan site, and participants will regularly communicate with each other there on our wall," notes Jason Graham of the bee research project Native Buzz. Participants use Facebook to share photos of the bee nests they monitor, articles of interest that they have found, and photos of bees or wasps in their gardens.

Scott Loarie, codirector of iNaturalist, a nature observation sharing site, says the site grew on the back of the photo sharing culture exemplified by sites like Flickr. "All we're saying is, instead of just sharing the photo, also share a little bit of information about it," such as the location and date where the photo was taken, he says. That way, what might otherwise be just a shared photo becomes a scientifically useful data point, with very little additional work required.

Many volunteers use their work within a citizen science project to forge connections outside of it as well. Loarie said that one iNaturalist contributor shared a number of nature photographs, and one of the frogs included was one that herpetologists had never seen before, which led to connections with those scientists as they sought more information. "It must have been a fantastic experience for an amateur, and the experts got a new species out of it," he says.

"It's amazing what monarchs can do to people," says Wendy Caldwell, Minnesota Larva Monitoring Project community program assistant. "This love inspires a lot of people to interact socially." While that interaction includes other project participants, as many volunteers monitor in groups with friends and neighbors, it also stretches far beyond the project's boundaries. Many participants put on workshops for youth groups, churches, garden clubs, or other local organizations.

This type of interpersonal connection is extremely valuable for people of all ages, from children to seniors. While the opportunity to participate in scientific research is a major initial draw for participants, the bonds that they form with people both within and beyond the project are a wonderful side benefit for everyone who takes part.

CHAPTER 4

Citizen Science Making a Difference

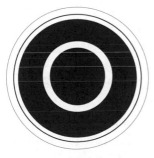

NE OF THE WONDERFUL THINGS ABOUT PARTICI-pating in citizen science is the opportunity to contribute to projects that truly help us understand the world we live in and protect it for ourselves and future generations. "I think that we could solve all the world's environmental problems by working together," says John Pickering, founder of Discover Life, which operates the Mothing citizen science project.

Many of the projects in this book are conducted on a large scale—nationwide or even worldwide. Keith Pardieck, US Coordinator of the North American Breeding Bird Survey, says participants welcome that opportunity. "A main selling point of the BBS is that each participant is directly contributing to national avian management and conservation efforts through local action," he says, adding that many of the species tracked by the survey aren't monitored by any other means.

Major projects such as these can have a huge impact. But research doesn't need to be global in order to be valuable. There are many citizen science efforts focused on small regions—and even dedicated to a specific site—that produce huge benefits for their communities.

Because these local projects cover relatively small areas, they don't need the huge base of participants that a national survey does to be effective. Individuals can have a larger impact on these projects as well. Where large-scale citizen science projects are generally led by professional scientists—often in conjunction with an advocacy group for the project's subject—several local projects have been founded and led by citizens, sometimes noncitizens who were simply passionate about a local natural resource.

The Sanctuary Education Awareness and Long Term Stewardship (SEALS) program is one example of local citizen science success. A partnership between the Gulf of the Farallones National Marine Sanctuary near San Francisco and the Farrallones Marine Sanctuary Association, the SEALS program was a reaction to alarming mortality rates of harbor seals at two sites at the sanctuary.

"The primary reason was the clamming season," says Mary Jane Schramm, media and public outreach specialist for the sanctuary. "Recreational clammers would frighten seals away, and often separate mothers from their pups."

SEALS volunteers collected environmental data and documented seal disturbances. In doing so, they discovered another significant issue: kayakers getting too close to seals. Because kayaks are quiet, Schramm says, their sudden arrival can mimic that of a predatory killer whale and frighten seals away, potentially leaving pups alone and helpless.

Thanks to that information, the sanctuary could take action. "We developed guidelines for paddlers and directed outreach among the paddling clubs," Schramm says. The program also trained volunteers to serve as interpretive guides at clamming sites, educating clammers about their potential impact and also suggesting clamming techniques that would not harm seals.

Those alerts had a big impact. "Most people don't want to harm wildlife," Schramm says. At one site, seal disturbances were reduced by 95 percent.

From 1996 to 2005, project volunteers donated time valued at more than $10,000 per year to the project. "That's money that the government did not have to spend," Schramm notes. Those volunteer efforts made a big impact. While the project ended in 2005, it ended on a successful note: seal pup populations were rebounding after fifteen years of declines. Schramm still sends annual press releases to keep the lessons learned fresh in the public's mind, and seal pup populations are still improving.

While SEALS is completed, the Beach Watch program still covers the Gulf of the Farallones, as well as Monterey Bay. That program provides long-term monitoring rather than addressing a specific issue, but it's still valuable when the sanctuaries face threats. When oil spills have occurred in the area, for example, program data have been used to accurately assess the damage, and have been used in court to win settlements for cleanup.

Other local projects contribute to conservation and habitat restoration decisions. "We have been very successful at cooperating with most of the land managers in the area," says Lee Ramsey, comanager of the Chicago-area Bird Conservation Network Survey. While the data that participants collect is useful for research, those connections with local land managers give the public the opportunity they haven't traditionally had to contribute to decisions that impact their community.

In many cases, citizens are the ones with the means to gather the information needed to make these decisions intelligently. "Decisions regarding impaired waters and how funds get spent all tend to happen at a high level of government," says Michele Tremblay, director of the Upper Merrimack Monitoring Program, which conducts water-quality testing on the Merrimack River in New Hampshire. "But government is pinched." As a result, almost all the data for the Merrimack now comes from volunteers.

"People feel like they're not so helpless," Tremblay observes. "They can see the health of the river in an up-to-date way."

"Back in the old days of resource management, the managers managed the resource based on their knowledge," says Bob DuBois, coordinator of the Wisconsin Odonata Survey. "Now the public understands it has a right to a seat at the decision-making table." Contributing to that process in a valuable way, he adds, requires a base level of knowledge, and citizen science projects give members of the public an opportunity to acquire it.

Locally focused citizen science projects are often good for finding specific but important information. The PRIDE Shorebird Survey monitors the Lower Santa Ynez River near Vandenberg Air Force Base in California. The river is important for shorebirds, because heavy rains there can expose mudflats that provide habitat for them. A bridge built in the 1940s reduced that mudflat habitat, and while the bridge washed out in 1969, berms from the bridge are still impacting the river. The survey's goals include determining the feasibility of removing the bridge remnants, as well as documenting conditions so that if they are removed, the impact can be assessed. "Only with fully considered protocols will a project meet with success by producing measurable results," says Tamarah Taaffe, treasurer of the La Purisima Audubon Society, which cohosts the survey with Point Reyes Bird Observatory Conservation Science.

Most national citizen science projects are led by either professional scientists as part of their research or advocacy groups with a long history of interest in the project's subject. Many local projects follow these leadership models as well. But the local scale is also where independent individuals can have a big impact by creating citizen science projects to address questions in their communities that larger projects don't focus on.

The Clark County (Washington) Amphibian Monitoring Project, for example, was started by Peter Ritson out of concern for the impact of the area's rapid urbanization on amphibians, as well as interest in the effects of the area's strong water quality regulations. "I am very involved and interested in protecting the natural environment—and a lot of people are," he says.

The project is a truly grassroots effort. The project's roughly sixty active participants determine the sites that are important for monitoring. "I think it's powerful that a community member can bring his fellow citizens to do this kind of work," Ritson says. In some cases, that community nature opens doors: a few landowners who wouldn't have let a representative of the state come on their property have welcomed survey volunteers, because those volunteers are also their neighbors.

Many project volunteers make presentations to neighborhood groups to help disseminate the information they collect. That's a fairly common activity for citizen scientists, even those with wider audiences. Eric Hanson, coordinator of the Vermont Loon Recovery Project, says that participants often share the information they learn, particularly with local lake associations. That sharing often leads to recruitment of volunteers to help with local lake monitoring.

Local citizen science projects can also be used to address questions of local interest. The Southwest Monarch Study is based in Arizona, and its central question is one that other areas of the country might not face: Where do Arizona's monarch butterflies migrate to? The migratory patterns of monarch butterflies both east and west of the Rocky Mountains are fairly well understood, but Arizona is south of the Rockies, and its monarch migrations are not so clear-cut.

As stated earlier, science is not solely the domain of professional scientists. If there is a local natural resource that you care about and want to preserve, or if you have a scientific question that existing research doesn't seem to answer, a citizen science project offers a way to gather the information you need while sharing the work with other people who have the same interests and concerns.

But while scientific expertise doesn't need to be the initial force behind a citizen science project, it is almost undoubtedly necessary if you want to achieve results that will stand up to scrutiny. Science does have a method that has served it well for centuries. Data collection protocols should be designed to minimize or eliminate inaccuracies and bias in the results, or the conclusions drawn from them may be unjustified and even harmful.

Experience can go a long way in helping to design data gathering procedures that are scientifically rigorous but still practicable by the members of the general public without a background in science. If you are looking for scientists who specialize in your subject and may be willing to help develop a citizen science experiment, consider nearby universities or associations or advocacy groups dedicated to your subject. You can also use similar projects in other regions or projects that examine a different but related subject as a model for your project.

Collaborative, grassroots science does require careful and respectful management of the people involved as well. "We realize that river systems have attachments to people with divergent interests," Taaffe says. "One can only endeavor to create a space where polarization is not encouraged." The results, however, can be uniquely rewarding to participants and make a big difference to the community as well. "We encourage community pride and stewardship by gathering data that can contribute to the management of important natural resources," Taaffe says.

Most citizen science projects exist to help protect and conserve natural resources. That's an area where locally focused actions can have a big impact, and these projects can encourage individuals and small groups to take positive action. Julia Parrish, executive director of the Coastal Observation and Seabird Survey Team project that monitors Pacific beaches, says that while some people seek out science out of interest in how the world works, others come to science out of concerns or worries they have about the world around them. "A program that can help create understanding gives more assurance and empowers people," she says.

Awakening individuals to the power they have to improve conditions can stretch beyond an individual citizen science effort. Jeremy Gieser, director of tours and marketing for Gastineau Guiding, says that one of the company's tours conducts citizen science monitoring of water quality in a creek near Mendenhall Glacier outside of Juneau Alaska. "Anybody can do that in their own backyard" with a simple kit, he says. "Hopefully this tour inspires people to get involved."

Building potential advocates and activists for the future is also a valuable effect of citizen science. The iNaturalist site, for example, welcomes the public to share their observations and photos of plants and animals. While those observations are scientifically useful, the site is also "accumulating a community of people active in certain areas of interests," says codirector Scott Loarie. Should the plants and animals of interest to site members face threats now or in the future, that community is ready to be mobilized to take positive action.

CHAPTER 5

Citizen Scientists at Work

HERE ARE HUNDREDS OF THOUSANDS OF PARTICI-
pants in citizen science throughout the country. Their reasons for participating are as diverse as they are—including passion for the outdoors, love of a specific animal, the opportunity to spend time with family, the personal educational value, and much more. Here are five of their stories.

Laura Molenaar

Laura Molenaar is a fifth-grade teacher in New London, Minnesota, and has been teaching monarch biology as part of her classes for years. As a teacher, she regularly participates in professional development classes at the University of Minnesota.

In 2001, Karen Oberhauser, director of the Monarch Larva Monitoring Project came to those classes to introduce the MLMP and pose some questions she and her students had developed. The project struck a chord with Molenaar, and that year she proposed it as a summer project for her students.

Ben Lange and Eli Kirkpatrick, two of Laura Molenaar's students, collect data for their group's contribution to the Monarch Larva Monitoring Project.

"I was looking for some students who were very engaged with science to do the project over the summer," she says. Six or seven students took part that first year, she recalls—the first members of a group and a project that has had surprising educational, social, and personal benefits.

Molenaar has now led student research for the past eleven years. Students make weekly observations for the MLMP at a land trust site a few miles away from the school that won't be built on in the future.

The kids are enthusiastic about the opportunity. "They're there at eight o'clock in the summer," Molenaar marvels. "As a teacher it's an affirmation of why you teach." The program's ranks have swelled to fifteen students, many of whom keep coming year after year. The group includes students from third grade through sophomores in high school.

That age range hasn't proven to be a problem. Instead, students have built relationships, forming a congenial peer group. "The seventh graders mentor fourth graders, and high schoolers mentor seventh and eighth graders," Molenaar says.

The group includes students at all levels of academic achievement as well. "Brighter students can take the project and run, but students who struggle academically also do well," Molenaar says. "They find a common ground."

The nature of the work does a lot to nurture relationships among students. "They're here for almost two hours," Molenaar says. "One hour is monitoring, but the other time is spent entering data, working on investigations, and just hanging out." That hanging-out time provides opportunities for kids to connect with one another, which is particularly important for those who are new or who do not have other activities where they can form that kind of bond.

Their shared interest in science and butterflies helps students break down borders that might be caused by differing ages, activities, or social groups. "At our end-of-season picnics, these are their best friends in the world," Molenaar says. And being with other kids who share their passion helps eliminate any worry that it might not be cool. "There's a lot of confidence in numbers," she says. "One of my students said, 'I love that we're nerds.'"

Molenaar's student-researchers have gained acclaim from outside their peer group as well. "It's great to see students get recognition when they don't get it because they can do something with a ball," Molenaar says. They have taken their results to science fairs, and students also went to an international monarch conference, where they saw how their data was being used. "That to them was such a plus, knowing what they're doing isn't just for play, it's real science," Molenaar says.

That reinforces the scientific experience—the act of following a protocol to collect and report data. For some, it has had long-term effects: former students, the first group of which are now college seniors, are going into science. One was so interested in the genetics of milkweed that he learned about through MLMP that he's now studying plant genetics at the University of Minnesota. And Molenaar has noticed firsthand how students have a deeper understanding of science in her classes after they take part in the summer program.

Not surprisingly, parents are pleased with the results. They've also supported the project by volunteering to drive students to the site. Molenaar says the school district has also been supportive, a relationship she helps to maintain by communicating results regularly.

Molenaar actually runs the program under the auspices of the district's community education program, which provides the insurance needed. She suggests that parents interested in starting this kind of group program investigate similar resources in their community. "You don't need a scientific background to be a success," she adds. "The kids will ask questions and you'll find the answers together. That's where the learning happens."

It's not, of course, necessary to have a group to participate in MLMP or most other citizen science programs. In Molenaar's case, however, it didn't hurt. "I would do this anyway because of my own personal interest," she says, "but the group adds a lot of fun and joy. If you could bottle the energy you get from a 12-year-old, there would be no energy crisis. It's infectious."

Cindy Burns

Cindy Burns has long had a strong interest in the natural world, and the ocean beach in particular—even though the ocean wasn't something she had access to growing up in Wisconsin. Her family had relatives in California, and those visits usually included a trip to the beach. "I was always fascinated" by those trips, Burns says.

She now works on the beach, managing habitat improvement for the snowy plover, a threatened bird species, near her home in Florence, Oregon, a town about sixty miles west of Eugene.

But even as a conservation professional, Burns wanted to expand her knowledge. In 2006, she says, "I was looking for something to volunteer for outside of my job, and I wanted to learn something new." She discovered the Coastal Observation and Seabird Survey Team (COASST) program when she read an article in her local paper about the program and how it trains volunteers to work in the field and identify beached seabirds. "It intrigued me, so I checked it out," she says.

Burns went to the training session and found the subject interesting. But there was also a serendipitous discovery: "I happened to sit next to the partner of someone who worked at the same office I did," she says. As a result, Burns was able to start volunteering with a built-in connection: COASST volunteers work in teams, and those groups form at the end of the training session. Burns and her coworker's partner joined the same group, along with four other volunteers, and began surveying a mile of Baker Beach north of Florence.

"At the time I didn't quite realize what I was getting into," Burns admits. "But I really enjoyed it and stuck with it."

Each of the group had some prior scientific training. Even so, COASST offers an opportunity for them to study the beach in an in-depth way that gives them opportunities to learn. "We're all continually learning new things about the beach ecosystems and the ocean," Burns says. Some of those members were retired, which gave them the time to attend environmental talks and then relate the information to the rest of the group. Other members had significant expertise with beach life and shared their knowledge generously.

The act of surveying also provides a robust educational experience. "There have been several species over the years that have washed up that I've never seen live," Burns says. "It's interesting to be able to see them up close." Among those discoveries was the carcass of an albatross, which many people never have the opportunity to see alive or dead.

While COASST focuses on seabirds, Burns notes that the project affords an opportunity to see animals well beyond that focus, including insects, whales, and live squid. "I enjoy being able to be out on the beach because there's something new happening each time," she says.

The work that Burns and her survey team do is still important for research and conservation purposes. But it's also an enjoyable social event. "It gets me out and meeting some like-minded people and contributing to Julia's [COASST Executive Director Julia Parrish's] database," Burns says. "It's fun spending time with good friends."

Attrition has reduced the survey group's size to three members. But families, including kids, often join in. In fact, a photo of Cindy and her children—identified as some of COASST's youngest volunteers—graces the project's website.

"It's an opportunity for the kids to be involved in a research project and interact with like-minded adults," she says. Naturally, the kids are able to help during the summer primarily, but they enjoy the experience enough to give their free time to it. "We include them as part of the team," Burns says. "They get to interact with the adults and truly feel as if they are providing a valuable service."

"It's a great thing for them to learn the scientific method and how to do careful measurements," she adds.

Burns's children are now both teenagers, and both join in. Her son has long been interested in birds and science, and her daughter enjoys spending time on the beach and associating with people who have similar interests. "This has been a great way for the kids to expand their knowledge."

Burns has shared the program with other kids as well. "For several years, I taught the basics about COASST during a watershed council summer camp for school kids" ranging from older elementary school through high school, she says.

She also welcomes the opportunity the project affords for her to work on something with her husband. "We've had a lot of good times making memories," she says. Burns likens it to an outdoor field trip they can take as a family every month: "It's sharing the learning process, having a good time together, and having a common interest with the kids."

Cox Arboretum Pollinator Volunteers

Cox Arboretum is one of several MetroParks in the Dayton, Ohio, area. Measuring almost two hundred acres, the arboretum includes woods, prairie, formal manicured gardens, and a native butterfly house.

Two years ago, the arboretum started developing a new garden around the butterfly house, the Pollinator Garden, which is filled with primarily native plants that attract bees and other pollinating insects.

The arboretum offers numerous volunteer opportunities throughout the park, and the Pollinator Garden is no exception. A group of volunteers tends the Pollinator Garden every week, led by Park Technician Meredith Cobb. "We weed our garden area, and collect and prepare, store, stratify when needed, and plant native species," explains volunteer Roberta Farinet. But in addition to their maintenance duties, the group takes a few minutes each week for science by collecting data for the Great Sunflower Project.

"I attended a training on bees by the Xerces Society [an invertebrate conservation organization] and learned about the Great Sunflower Project there," Cobb says. "I thought it would be a great way to get the volunteers engaged and involved with pollinators" while affording the volunteers an exciting opportunity to learn and take part in scientific research.

"I was attracted to partaking in the Great Sunflower Project in correlation to why I volunteer—to learn and promote conservation and life cycles of pollinators," says group member Sam Fullen. "The GSP collects data directly pertaining to pollination numbers, so it was logical to spend a brief fifteen minutes a week to try to assist in that data collection."

The Great Sunflower Project, which entails recording the activity of bees on specific viable plants, is just one part of what the volunteers do. But it's an enjoyable part. Fullen says the project offers an excellent opportunity to learn about the plants they tend and the pollinating insects that visit them, and the opportunity to contribute to a national research study is valuable as well. "It's also nice to take a short break from weeding," he adds.

About five to ten people volunteer each week. "Most of the people involved are gardeners, or people who are interested in native plants and gardening for wildlife," Cobb says. "Some are beekeepers too—we have some native bee houses in the garden as well."

Farinet, for example, has native plants, host plants for butterflies, and nectar-producing plants for pollinators in her home garden. "But it is a small space," she says. "I joined the Pollinator Garden Group at Cox to show the public what can be done in their yards."

Participating as a group is a useful experience. "I prefer to work with a group on this because data collection is simpler and there is direction and help when needed," Fullen says. "As someone with less plant knowledge than other, more experienced volunteers, others help me to know exactly what should be monitored, or how to tell the difference between the different kinds of bees."

"Doing it as a group keeps us consistent, and makes it a broader experience sharing observations among us," Cobb adds.

Family members—sometimes including children—join in occasionally, and Cobb says she hopes to add more volunteers and more public participation in future years.

Volunteering serves as a social outlet for participants, an excellent learning opportunity, and a chance to work to improve habitat needed by bees and other pollinating insects. "For me it is emotionally wonderful, keeping my frustration with other parts of my life within reasonable limits," Farinet says.

The Great Sunflower Project is a welcome addition to the arboretum's work, blending with its mission and providing a worthwhile outlet for volunteers to give their time and labor. "I am very interested in integrating science within the arboretum, and for the visitors here," Cobb says. "By participating in the GSP we are collecting baseline data set and will be able to see if things change in the future. It's also just fun!"

Ken and Zoe Strothkamp

Ken Strothkamp has a PhD in biochemistry, and he teaches and conducts research at Lewis & Clark College in Portland, Oregon. But he's also active, with his daughter, in citizen science, and the two sides have even merged in surprising ways, as Strothkamp has begun his own scientific inquiry and built a far-flung network of contributors to help carry it out.

It started several years ago. Someone gave his daughter Zoe, then 4 or 5 years old, a butterfly net, and she caught a large moth. "I knew nothing about butterflies or moths at the time," Strothkamp says. "but I didn't want to look like an idiot to my kid."

In his attempt to discover the animal's identity, Strothkamp found the Butterflies and Moths of North America website (www.butterfliesandmoths.org).

But the site had no record for the specimen from his county in Oregon, so he submitted photos. Before long a lepidopterist had confirmed the moth's identity and added the record of Zoe's find to BAMONA's database. "My daughter was thrilled to learn she had contributed this bit of data," Strothkamp says. He was even able to help her visualize her contribution: by capturing before-and-after species maps from the site, they could see the county that got colored in as part of the moth's range because of Zoe's find.

She didn't stop. Zoe and Ken kept finding, photographing, and releasing butterflies and moths. "She found about a dozen new county records for Oregon," Strothkamp says, and ultimately was featured in the local newspaper for her efforts.

And then, one of her finds, a spotted tussock moth, laid eggs that hatched into fuzzy, colorful caterpillars. "She had lots of questions and I had no answers," Strothkamp recalls. This time, however, Internet research didn't lead to clear answers—the information and particularly the photographs Strothkamp found conflicted with each other.

In fact, the conflicting photos weren't wrong. The spotted tussock moth has a huge range, about thirty degrees of latitude, from just north of the Mexican border to Yellowknife in Canada's Northwest Territories, and from coast to coast. The species has significant variations in appearance throughout its range. Curiosity fed itself, and eventually developed into what might be considered a citizen science project in its own right, as Strothkamp has begun studying the reasons for these variations.

"It is my hope to relate the evolutionary history of the species to the environmental changes in North America and show how the species responded to, and was shaped by, the significant climate changes that occurred since the last glacial maximum," Strothkamp says.

To do that, he needs observations and specimens from across the country. Through sources like BugGuide.net and *The Journal of the Lepidopterists' Society*, Strothkamp has found a "small army" of more than fifty people who will share observations, photos, and specimens of the spotted tussock moth with him.

Here, his formal training and his access to a fully equipped lab are critical, as he works to analyze and identify the pigments responsible for the colors and color variations of the caterpillars and moths. But it also wouldn't be possible without his contributors, which include casual observers and serious amateur naturalists, who can cover the geographic range that Strothkamp couldn't on his own.

Ken and Zoe Strothkamp examine a spotted tussock moth.

"So much of our understanding of living things is so limited. Professional scientists don't have the time to go out and scour every inch of the world," Strothkamp says.

The research has practical applications as well. In the southern part of its range, the moth lives only at high elevation, where the climate is more similar to the north. As the climate warms, those communities, which are already isolated, may disappear quickly. "I'm hoping if we can determine how the organism has evolved, we can see how it copes with modern-day climate change," Strothkamp says.

Doing so may provide clues to how other species will adapt as well. "Often you start out very specific and make discoveries with more general applicability," he says.

Strothkamp provides annual updates to contributors. "I make it clear to them that the goal is to make a contribution to science, and ultimately to publish in the peer-reviewed literature." It's a relatively slow process, due to the moth's life cycle (it produces only one generation per year, and flies for only a couple months of the year). Strothkamp has published one paper about the research so far, and plans more. Contributors are cited as coauthors or acknowledged by name, depending on their involvement.

"I would like to leave behind a record of some little bit of knowledge that I've discovered," Strothkamp says. "I think a lot of people are motivated by this."

And Zoe, whose initial interest sparked the project, is still motivated to explore as well. "Her world is broader now, but she still remembers," Strothkamp says. "It had a significant impact on her at a very early age."

Want to join Strothkamp's investigation into the evolution of the spotted tussock moth? He is always looking for contributors; e-mail him at kenstrothkamp@ lclark.edu.

Mike Newkirk

Mike Newkirk's passion for the common loon dates back to his childhood— even though he didn't quite realize it then. He grew up on a small lake near Albany, New York. He says the lake was nothing but green muck in the summer, but he and his friends gladly hung out there anyway.

But when he was 12, he discovered that lakes didn't have to be green muck. That was the year he went to camp in the Adirondacks. "I was taken with the fact that I could stand in the lake up to my chest and see my feet," Newkirk says.

As an adult, he visited the New York State Museum, where an exhibit reminded him vividly of that lake. But when he returned to the Adirondacks, he realized it wasn't the same. Acid rain had thinned the fish population, which had forced the loons to move on, and their distinctive cry was what he missed.

"That put me on a mission to hear loons again," Newkirk says.

He made trips to Vermont to try to hear the call, and eventually moved to the state, but his quest remained unsuccessful until he saw an article about the Vermont Loon Recovery Project's LoonWatch census. He spent the day at one of the sites where loons had been known to nest and, once again, heard nothing.

That adventure ended well, however, because the next day as he was preparing to leave the lake the loons finally did call for him.

Newkirk still sought the loon after that first hearing. He found a list of where loons were and started making camping trips to those sites. And when the opportunity arose to create nesting floats for the VLRP, he took responsibility for the Chittenden Reservoir.

Mike and Emilia Newkirk anchor a loon nesting float in position.

The reservoir had no loons on it at the time, but Newkirk and his son built the raft in hopes of attracting a pair by simulating the shoreline habitats where loons build their nests.

In the second or third year, Newkirk recalls, heavy storms broke the raft loose from its designated location. He and his son eventually found it on the other side of the reservoir—with a loon's nest on it. The nest failed that year, but two chicks were born on the site the next year, and the nest produced young for several years after that.

Newkirk has four children, the youngest of whom just started college, but all four have joined him in the project. "It really got my family and the kids involved in the outdoors," he says. One of the kids even got a loon tattooed on her wrist. Another wrote a paper about loons in high school that earned him a $1,000 grant.

The kids are all still passionate about the outdoors, which Newkirk attributes in part to their experiences with the loons. The family also made nature an important part of its vacations. "We had one rule of thumb: we had to take a vacation involved with nature once a year," Newkirk says. "It puts a higher expectation on the trip—they're going somewhere to have an adventure instead of sitting in a casino somewhere."

And thanks in part to this shared passion, "we are still as connected as we were when they were kids," Newkirk says.

This interest in loons has extended beyond the immediate family as well. Newkirk's brother and his family in the Catskills, and his wife's sister and her family in Dallas, have all been introduced to the loon and taken a strong interest. Newkirk has also met people at the reservoir who have shared his interest and who helped monitor the float when the family couldn't get there.

Those times were rare, however. The family didn't take its commitment to the project lightly; no one was allowed to simply decide they had better things to do. Bad weather was also no excuse. The family's last trip to monitor the raft took place during Mother's Day, but the day had terrible rainstorms. Newkirk acknowledges that the family was tempted by the warm, dry restaurants they passed on the way to the reservoir, but they continued on regardless. Even though the family had to spend several hours in torrential rains to make major repairs to the float, it proved to be a rewarding experience. "We were too wet and dirty" to stop at the restaurants on the way home, "but we had a great sense of accomplishment and had our usual Sunday night family dinner knowing the task was done for the season," he says.

Newkirk no longer maintains the nesting float, as it required a boat for access; he sold his when his kids grew up. He does still take part in the Loon-Watch aspect of the VLRP, though. And the imprint of the project on him and his family remains strong. Newkirk credits the family's work, in conditions good and bad, as a strong contributor to the tight bonds they have forged. As he says: "It forced us to be a good family."

CHAPTER 6

Getting Started at Your Library

HE GUIDE TO CITIZEN SCIENCE PROJECTS IN THIS book is intended to give you a good start in finding the project that is right for you and your family based on your interests, location, scientific expertise, and the amount of time you want to devote. Each project's website can give you much more detail about what the project entails and the scientific research you'll be contributing to, and many offer more background information about the subjects they study as well.

But there is another great resource for you to take advantage of as you enter the world of citizen science, or when you've been participating in citizen science for a while and are ready to expand your knowledge: your local library.

Obviously, libraries offer books for borrowing at no charge with a library card. Each section of the projects guide includes a listing of books recommended by scientists and librarians to help you learn about related topics. Your library's catalog (which is usually accessible in the building or online) can tell you which of these books your library offers. The librarian can help you find more related books, and can usually get books that the library doesn't carry through interlibrary loan services.

Librarians are outstanding guides to online information as well. Most libraries offer free access to the Internet, and more than 70 percent offer a community's only free Internet access point, according to a 2010 Harris Interactive poll. "Libraries of all types have always had programs that build information literacy skills for students and patrons," according to the American Library Association's *State of America's Libraries, 2012.* "As technologies have changed and influenced how people search for, find, and use information, libraries have adapted their programs to ensure that their users have the requisite skills—both technical and cognitive—to be able to take advantage of the resources and opportunities afforded by the internet and the digitization of information." An important skill librarians can offer is searching the "Deep Web"—the millions of web pages that search engines like Google or Yahoo don't index and cannot retrieve. The Deep Web is estimated to be several orders of magnitude larger than the easily explored "Surface Web."

Libraries also often have subscriptions to databases of articles on specific, often specialized topics. These databases often contain information that simply aren't available publicly through the Internet. The librarian can tell you what databases the library has available and how best to search them for the information you're trying to find.

Most libraries host presentations from a wide range of speakers. Many of the coordinators of the citizen science projects featured in this book have spoken at their local library—or even beyond. For example, Kurt Mead, leader of the Minnesota Odonata (dragonfly) Survey Project, spoke at about two dozen libraries throughout Minnesota in 2005 alone, when many libraries in the state held summer reading programs with the theme "Buzzing about Bugs." He has continued presenting at libraries about the dragonflies and damselflies his project surveys steadily since then.

Gastineau Guiding, a scenic tour company in Juneau, Alaska, conducts training for its guides for the company's citizen science tours at local public libraries, says Jeremy Gieser, the company's director of tours and marketing. But those training sessions are open to the public as well, so they can serve as educational programs about the whales, algae, and water quality that the tours survey. In March 2012, Westport Public Library in Connecticut hosted a presentation by Yale astronomer Brooke Simmons, who spoke about her research on supermassive black holes as well as the importance of citizen scientists in analyzing data as part of the Galaxy Zoo astronomy program.

These library programs can lead to delightful coincidences. Gail Morris makes several presentations at libraries about the Southwest Monarch Study and about monarch butterflies more generally for both adults and kids. In 2011, one of the monarchs Morris had tagged and released was recovered and reported by a person who had attended one of her library talks just a month and a half prior.

At their most basic level, libraries are community spaces. Most have private rooms set aside for a variety of community uses. For that reason, many citizen science projects hold training sessions at libraries. It depends on the project, but some of these training sessions are also open to interested members of the public, so you might be able to sit in and see if the project is right for you while meeting other people who share your interests in science. You can usually find what training sessions might be happening on your local library's calendar of events.

Libraries have always taken their educational mission seriously. But now, many public, school, and college libraries are taking particular interest in STEM (science, technology, engineering, and math) activities. "Libraries and museums are improving learning in science, technology, engineering and math, a national priority for US competitiveness," according to the Institute for Museum and Library Services. Some are introducing hacker spaces with a variety of equipment where patrons can experiment with their own creations. Others have designed facilities with dedicated outdoor learning spaces that could easily facilitate the kind of observations that many citizen science projects are built upon. Almost all libraries have the computer technology that makes contributing data to a citizen science project easy. And many are located on or near public parks or nature preserves teeming with life that scientists would love to document.

Greensboro (North Carolina) Public Library's Kathleen Clay Edwards Family Branch is one such library. The library is located in the ninety-eight-acre Price Park, and its collections and programs have an environmental focus, says environmental resources librarian Melanie Buckingham.

In 2010, the library partnered with numerous organizations, including the Piedmont Bird Club, Greensboro Parks and Recreation, local universities, and the Audubon Society to conduct a bioblitz in the park. "Our goal was to get a baseline of what's here in Price Park and to think about ways to protect it," Buckingham says. The library helped to publicize the event to the public, and its Geographic Information Services department helped to support the project's mapping efforts.

"It makes sense to use the park as a living classroom," Buckingham says. "If patrons learn to see what's out there and to appreciate them, then when things are imperiled, they will get involved."

Like many libraries, the Greensboro Public Library has a bookmobile and other outreach efforts to provide service to people who might not otherwise be able to easily access library services. The library plans to retrofit the bookmobile to create a mobile environmental classroom.

Recently, the library started an environmental sciences service learning project for high school students, which will also serve as a test case for programs with families. The library has also made initial contact with the National Phenology Network to get involved in its Nature's Notebook citizen science program.

Celebrate Urban Birds Coordinator Karen Purcell says that libraries are among the many community organizations that Celebrate Urban Birds partners with, hosting a wide range of programs ranging from webinars and Skype-based presentations to summer-long programs and major community events. "The libraries seem to be a great place to be able to have an event," she says. Libraries that have participated include Santa Clara City Library in California, which hosted a Celebrate Urban Birds booth at the city's Arbor Day/ Earth Day celebration where children could learn how to take part in the program and make bird puppets; the Easttown Library in Berwyn, Pennsylvania, which created a display on birds and birding and organized a bird walk and a bird-themed family movie night; and the Windsor Park Library in Austin, Texas, which held a community, arts, and nature festival centered on birds.

It's difficult to quantify how many libraries are active participants in citizen science projects by creating library programs around citizen science activities. My research has turned up few examples of formal connections, although it's entirely possible that participants do use libraries in this way without alerting the leaders of the projects. Regardless, given the factors above, I think it's only a matter of time until libraries do become formal centers for citizen science.

Many citizen science projects would be easy to adapt as library programs. Projects such as School of Ants, which requires about an hour of time and simple materials to collect samples of ants from a site; the Community Collaborative Rain, Hail, and Snow Network, which tracks precipitation at individual sites nationwide; or the Zooniverse projects, which comprise simple, computer-based analyses of astronomical, biological, or historical data, are all simple to explain, suitable for small groups to participate in, and able to be completed in the amount of time a library program would allow for.

Other, more in-depth citizen science projects might not be suitable to adapt data collection sessions as library programs. But libraries still have the space, resources, and expertise to provide value for participants, either individually or as gathering spaces for groups of participants who want to come together to share information and learn from each other.

It's impossible to say precisely what the library in your city, county, school, or university might offer that can help you in your citizen science efforts. Libraries are truly local institutions, accountable to their specific audiences, so one city's library may offer very different services from the library in a neighboring city.

But that lack of uniformity offers the benefit of flexibility. Since your local library is accountable to the people of your neighborhood, rather than the state or some national organization, there's a good chance that you can help influence the materials and services they offer. Most libraries welcome suggestions about the books and other resources they should purchase as well as the programs they should offer. Many also welcome patrons to volunteer to share their knowledge and their passions. If you are interested in making your library a hub for citizen science for yourself and your community, there is plenty of opportunity for you to make it happen.

Not sure where your nearest library is? The website atyourlibrary.org offers a function to search by zip code to find a library location near you.

A DIRECTORY OF CITIZEN SCIENCE PROJECTS

HERE ARE AN AMAZING VARIETY OF CITIZEN SCI-ence projects available. They cover dozens of different topics, and each requires different levels of knowledge and different time commitments. The following directory is intended to help you find the project that fits your interests and abilities.

Projects are arranged alphabetically within topical categories. Most projects that are dedicated to a specific type of habitat can be found by searching for the plants and animals that live there. The exception is those projects covering beaches, rivers, and wetlands, which often collect a broad range of data that don't fit neatly into a single category. Those projects can be found in separate categories. Each project description outlines the type of work and commitment level that participation requires.

Most citizen science projects are open to all ages. But many are particularly well suited to younger participants. These are noted at the end of each listing. Listings also specify which projects are limited by time of year or region, if there is any fee to participate, and the project's website.

Analytical Games and Puzzles

Most citizen science projects involve fieldwork to collect data. But that's not the only way the public can contribute to research. Several projects have huge quantities of data collected, often in photographic form that is difficult for computers to analyze quickly and effectively. Others are looking to solve puzzles that computers can model—but not perfectly, leaving room for human intuition to make improvements. Many of these projects have adapted their central problems into casual games that anyone can play. With these projects, a few minutes on a coffee break can help turn mounds of information into understanding.

Several projects in other categories include this type of activity. See also Whale Song Project (page 115), Galaxy Zoo (page 147), Moon Mappers (page 150), and Stardust@home (page 151).

EteRNA
http://eterna.cmu.edu

EteRNA is a puzzle game in which players design models of RNA, which carries genetic information that directs the creation of proteins. Players create RNA strands using the four nucleotide bases in order to make stable base pairs and targeted shapes.

"Recent discoveries have shown that RNA can do amazing things," the project's website says. "By playing EteRNA, you will participate in creating the first large-scale library of synthetic RNA designs." Researchers hope that among those designs will be ones that can help to fight a range of viruses.

After learning the game and learning its rules, which are based on how the nucleotides that make up RNA bond, experienced players can turn their attention to special lab puzzles. When they come up with solutions to these puzzles, project researchers will attempt to synthesize their model solutions in real life.

♟ AGE RANGE All ages   **▦ TIME FRAME** Year-round   **⊕ REGION** Worldwide   **$ FEE** None

Foldit

www.fold.it

Foldit is a downloadable puzzle game based on protein folding. Proteins are made up of a long chain of amino acids, but rather than staying in that chain, they fold over onto themselves into compressed shapes that specify the protein's function.

ANALYTICAL GAMES AND PUZZLES
—

"Proteins are the workhorses in every cell of every living thing," the game's website explains. But while proteins are critical to human life, some proteins help bacteria and viruses cause disease. Understanding their structure can help scientists design new proteins to counter those harmful proteins.

The Foldit project is currently trying to determine if humans can be more effective than computers at predicting protein structures. If that experiment is successful, the eventual goals of Foldit include having players map the structure of proteins that aren't known and to design new proteins for specific tasks.

The game consists of puzzles that participants try to fold as compactly as possible while obeying rules about how various atoms in their structure behave toward one another. Players are given a 3-D model of a protein and use a computer mouse to adjust sections of the model to form the most efficient shape. Puzzles are scored based on how successful the shape is.

Foldit was developed by the Center for Game Science at the University of Washington and the University of Washington Department of Biochemistry.

AGE RANGE All ages **TIME FRAME** Year-round **REGION** Worldwide **FEE** None

---

Mapper

www.getmapper.com

Mapper is an online, gamelike project from NASA in which participants analyze photos of the bottoms of Pavilion Lake and Kelly Lake in British Columbia in hopes of helping to find life on other planets.

The lake bottoms contain microbialites, which are rocks whose creation was shaped by the bacteria that lived on their surfaces. Microbialites are now rare on Earth, but they were the only life on the planet for its first billion years. Since they formed in the harsh environment of early Earth, scientists hope that studying them can help them to determine the types of environments in outer space where life might form.

Hundreds of thousands of images of the lake bottoms have been taken. Mapper enlists the public's help in tagging those images so researchers have a better idea of what features can be found where and which features warrant further study.

Participants must pass a tutorial and quiz before being able to tag photos. They earn points for completing the tagging of photosets, which can earn places on the leaderboard. Points also unlock downloadable wallpapers, and after tagging enough photos, participants can unlock advanced classification activities.

AGE RANGE All ages　**TIME FRAME** Year-round　**REGION** Worldwide　**FEE** None

---

Phylo

http://phylo.cs.mcgill.ca

Phylo is an online puzzle game from McGill University in Montreal that seeks to examine how DNA and RNA sequences can be arranged. Players arrange several rows of colored boxes, representing sequences of DNA, as well as gaps that represent mutations. Points are awarded for matching boxes in adjacent

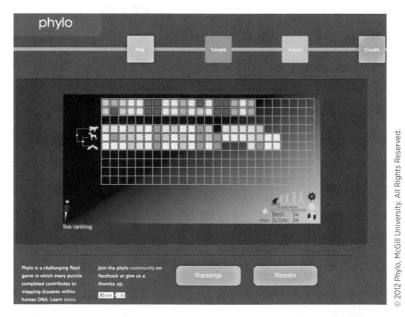

A DNA-sequencing puzzle from Phylo.

rows. Gaps in rows cost points, but may help increase scores by making more matches possible.

This casual box-shifting puzzle models a huge scientific question. "One of the most important problems in computational biology is comparing the DNA of species," says Jérôme Waldispühl, coleader of the Phylo project. By comparing, he adds, researchers can identify regions that are identical in the two genomes. While significant parts of DNA do not code for any proteins and have unknown biological functions, the sections that are identical across species likely provide something critical for life. Researchers hope finding them can help determine the source of some genetic diseases and demonstrate shared evolutionary origins.

Computer programs can perform the kind of alignments that players do well—but not perfectly. Those imperfections are where human intuition can help. "We generate puzzles using an alignment made by the computer and finding small regions that haven't been well aligned," Waldispühl says. Results have been encouraging so far: In 739 regions, human players improved the computer alignment in 70 percent of cases.

ANALYTICAL GAMES AND PUZZLES
—

AGE RANGE All ages   **TIME FRAME** Year-round   **REGION** Worldwide   **$ FEE** None

---

Valley of the Khans Project   http://exploration.nationalgeographic.com

National Geographic's Valley of the Khans project is an online hunt for Genghis Khan's tomb, which remains undiscovered nearly eight hundred years after his death, with few records that provide clues to where it might be.

Local traditions preclude a traditional archaeological dig. The project offers a way for citizen volunteers to help guide noninvasive exploration of the land by examining high-resolution satellite images.

The technique was inspired in part by the search for explorer Steve Fossett, who was flying a plane that disappeared over Nevada in 2007. More than fifty thousand volunteers scoured satellite images on Amazon's Mechanical Turk website in an attempt to locate the crash site. "It was an aha moment," says Albert Lin, the project's principal investigator, where he realized that volunteers would be willing and able to examine satellite images for archaeological features in Mongolia as well.

"We wanted to call upon the collective of human perception and intuition to determine what's unusual" in those pictures, Lin says.

After a short tutorial, participants tag satellite images of Mongolia for features including roads, rivers, modern buildings, and ancient structures such as burial mounds and other archaeological sites. After tagging, participants can compare what they saw to what other people tagged in the image.

The research team has collected millions of inputs so far. Each image is examined by hundreds of users independently; sites that most users agreed were of interest helped to guide the researcher's choices of where to make archaeological trips. The team has honed in on one location that holds special promise, but Lin says the site is still collecting data and may well be adapted for other archaeological projects in the future.

The project is open to anyone and suitable for all ages. Lin says that feedback has been positive, with users reporting that the experience has inspired them to take archaeology classes to continue learning about cultural heritage.

AGE RANGE All ages **TIME FRAME** Year-round **REGION** Worldwide **$ FEE** None

Zooniverse
www.zooniverse.org

Zooniverse is a suite of projects that crowdsource the analysis of huge amounts of data. Most Zooniverse projects study topics in astronomy, although some examine climate, archaeology, and nature as well.

Zooniverse projects include:

Ancient Lives, in which participants help transcribe papyri from Greco-Roman Egypt

Bat Detective, a project to help classify bat calls as a mechanism for monitoring bat populations

Cell Slider, in which users help classify tissue samples to analyze cancer data

Cyclone Center, which classifies historic satellite images of tropical storms to help improve understanding of cyclones

Galaxy Zoo, which analyzes images from the Hubble Space Telescope to investigate how galaxies form (see a full description on page 147)

The Milky Way Project, which investigates infrared images from the Spritzer Space Telescope to find bubbles in space where stars might be forming

Moon Zoo, which classifies images from NASA's Lunar Reconnaissance Orbiter to help researchers study the moon's surface

Old Weather, in which participants help transcribe weather observations made by Royal Navy ships around the time of World War I to contribute to climate model projections

Planet Four, a study of wind patterns on the surface of Mars

Planet Hunters, which looks for exosolar planets by examining brightness measurements of more than 150,000 stars taken every thirty minutes to identify dips that might indicate a planet's transit

Seafloor Explorer, which is creating a library of seafloor life along the northeastern United States

Solar Stormwatch, which investigates how solar storms affect conditions in space and on Earth

The Whale Song Project, which seeks to better understand the various calls made by whales (see a full description on page 115)

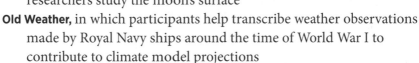

AGE RANGE All ages **TIME FRAME** Year-round **REGION** Worldwide **FEE** None

---

Animals

AMPHIBIANS AND REPTILES

Amphibians and reptiles are distinct classes of animals, but they are often studied together under the classification of herpetology. Amphibians in particular are often important indicators of ecological health; their lifespan includes both aquatic and terrestrial phases, so problems in either habitat will affect them, and they have permeable skins that make them particularly susceptible to pollution. As a result, the number of citizen science projects to track and monitor them is hardly a surprise.

Note that there are efforts in many states to create herpetofaunal atlases that show what types of reptiles and amphibian can be found in the state and where. Only those that are still being built are included in this guide; however, many of the more well-established state atlases do still welcome volunteer contributions. An excellent guide to these projects is hosted by the Kansas Herpetofaunal Atlas at webcat.fhsu.edu/ksfauna/herps/index.asp?page=links.

FROGS

Calling Frog Survey

www.habitatproject.org/frogsurvey

The Chicago Wilderness Habitat Project's Calling Frog Survey seeks to detect amphibian population changes in the Chicago region. "In the 1990s, scientists throughout the world got together at conferences and noticed many were finding there weren't as many frogs as there used to be," says Karen Glennemeier, Habitat Project conservation scientist. But those observations, she adds, were all anecdotal rather than numeric. A monitoring program was needed to determine if those decreases were real, and if so, if they were due to natural fluctuation or long-term trends. This need sparked scores of frog monitoring projects throughout the world, including the CW Calling Frog Survey.

The project has a relatively low time and experience requirement, and is suitable for kids to participate with their parents. Participants need to learn the calls of thirteen frogs and toads, and they must also commit to three site visits per spring, each lasting one to two hours, and attendance at a two-hour workshop in late winter. Participants are expected to continue with the project for multiple years.

"I think the thing that keeps people coming back is how cool it is," Glennemeier says. Project volunteers conduct surveys in the evenings, when few if any people are on the sites. "You feel like you're in a private, magical world," she says.

The data allows researchers to examine broad patterns and trends. The cricket frog, for example, was once the most common amphibian in Illinois.

It nearly disappeared from the northern part of the state by the time the survey began in 2000, but Glennemeier says data indicate stability in remaining populations. The data participants collect now "give us a backdrop against which to compare future numbers," she says.

It also gives participants, who already care about conservation and nature, the tools to advocate effectively. One monitor in northwestern Indiana discovered that a park district's planned restoration would have destroyed a depression used by many amphibians for breeding. That monitor was able to persuade the park district to change its plans. "That was a wonderful little victory that happened because she was out there paying attention and had the wherewithal to have a conversation" with decision makers, Glennemeier says.

AGE RANGE All ages **TIME FRAME** Spring **REGION** Northern Illinois, northwest Indiana **FEE** None

---

FrogWatch   www.frogwatch.ca

FrogWatch is an initiative of NatureWatch that monitors frog populations throughout Canada. Because frogs and other amphibians require both land and water, they are particularly sensitive to changes in the environment and are a good indicator of the environment's health. "[The project] helps to situate local changes in a broader perspective," says Marlene Doyle, NatureWatch manager. All the data collected by the project is available freely on its website.

Participants monitor frog and toad calls at a location of their choosing for at least three minutes around dusk during local breeding seasons. Participants are encouraged to monitor at least weekly, although even single reports are welcomed. Participants report the number and type of calls heard, as well as air and water temperature data. More than twelve thousand observations have been contributed to the project so far.

"Many people are quite excited and interested in the frogs they see," Doyle says. "They want to see if it's normal."

AGE RANGE All ages **TIME FRAME** March through August; some provinces may have shorter seasons **REGION** Canada **FEE** None

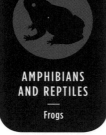

FrogWatch USA

www.aza.org/frogwatch

Founded in 1998, FrogWatch USA is a citizen science project of the Association of Zoos and Aquariums, with the aims of conserving amphibians and helping participants learn about the wetlands in their communities.

"Due to their semipermeable skin and life histories that include time spent in both aquatic and terrestrial environments, frogs, toads, and other amphibians are sensitive to a variety of ecological changes that may be otherwise difficult to detect," according to the Association of Zoos and Aquariums. "Therefore, they can serve as indicators of environmental health and may help detect changes that could ultimately affect human health." State agencies, land managers, universities, and museums and science centers use FrogWatch USA data.

Participants visit wetlands several times during the frog and toad breeding season to identify the species present by their calls, as well as weather information and the intensity of the calls. Surveys take place in the evenings shortly after sunset and require two minutes of silence to allow amphibians to acclimate to the presence of humans, followed by three minutes of observation. Participants select how often to monitor their sites, although the project recommends monitoring at least twice a week.

AZA says that FrogWatch USA is well-suited for families, groups, or individuals. "Participants are encouraged to monitor at wetland sites that are convenient for them" and require no equipment beyond a data sheet, thermometer, pen or pencil, and clock or stopwatch, according to the association.

More than one hundred local FrogWatch USA chapter coordinators recruit and train volunteers. These chapters are generally affiliated with zoos, museums, or similar organizations. Many local chapters offer opportunities such as group surveys and guest speakers throughout the season. Participants without a nearby local chapter can also participate in the program's online community.

AGE RANGE All ages **TIME FRAME** February through August **REGION** US
FEE None

Global Amphibian BioBlitz /
Global Reptile BioBlitz

www.inaturalist.org/projects/global-amphibian-bioblitz
www.inaturalist.org/projects/global-reptile-bioblitz

AMPHIBIANS AND REPTILES

Frogs

The Global Amphibian BioBlitz and Global Reptile BioBlitz are sister projects hosted on iNaturalist, a network where citizen scientists can share their observations of the natural world. The projects seek to document all the world's amphibian and reptile species—significant numbers of which are threatened by climate change and increased land usage by humans.

"Both of the BioBlitzes are partnered with the IUCN Red List" of threatened species, says Scott Loarie, administrator for both projects. IUCN, the International Union for Conservation of Nature, is an international group responsible for assessing the conservation status of species.

Contributions to the BioBlitzes help to improve IUCN's assessments, because there are many species that science simply doesn't know a lot about. "Just a few data points can do a lot to inform those conservation decisions," Loarie says. The BioBlitzes also share information with the Encyclopedia of Life, an online species database, and the Global Biodiversity Information Facility, an online database of biological specimens held by museums.

Using a free iNaturalist account, participants in these BioBlitzes post photos of their sightings, as well as the sighting's date, location, and the best identification possible. Experts will help identify unknown photos. Participation is open to all ages, but the iNaturalist account holder must be at least 13.

So far, the projects have documented more than 1,300 of the world's 6,830 amphibian species, and more than 1,200 of the world's 9,413 reptiles. A significant number are endangered, and at least one, Holdridge's toad, had not been seen since 1986 and had been believed extinct. In other cases, the BioBlitzes have significantly improved the available samples. "Many species we have no pictures of, or they may have been collected decades ago and only exist as a wrinkled specimen in a museum," Loarie says. "So just having a picture is very valuable."

AGE RANGE All ages   **TIME FRAME** Year-round   **REGION** Worldwide   **FEE** None

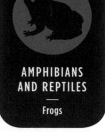

Kansas Anuran Monitoring Program

www.cnah.org/kamp

Established in 1998 by the Kansas Department of Wildlife and Parks, KAMP monitors Kansas's twenty-two species of frogs and toads through surveys conducted by citizen scientists at permanent roadside routes. The information will help to determine status and population trends of the species, as well as habitat quality. "Their presence, absence, and most notably disappearance from an area may provide information on the condition of Kansas' wetland habitats," the project's website says.

Participants survey their routes several times during the calling season. Surveys generally require a couple of hours to conduct, and take place in the evenings when frogs are most active in their calls. Each route has ten stops, and participants listen for five minutes at each stop to determine which frogs and toads are calling.

In addition to detailing the species of frogs and toads that are vocalizing, surveyors record wind and sky conditions and the relative quantity of calls.

AGE RANGE All ages   **TIME FRAME** February through August   **REGION** Kansas   **FEE** None

North American Amphibian Monitoring Program

www.pwrc.usgs.gov/naamp

The NAAMP assesses frog and toad population trends. Citizen scientists identify and count local amphibians by their breeding calls at assigned roadside routes, each of which has ten stops. Routes are assigned by state coordinators. Volunteers listen for five minutes at each stop to record the species heard, the intensity of their calls, and environmental data.

"People care about the health of the natural environment around them," says Krista Larson, state coordinator for Minnesota. "Many people report seeing and hearing fewer frogs today than when they were growing up, and they want to do their part" to help reverse the trend.

Participants survey the route three times per year on spring and summer evenings, and are expected to participate for multiple years. Volunteers must learn to identify local frog and toad calls and must pass an online test before their data can be used in national analyses.

A single volunteer collects data, although assistants are permitted to accompany the observer on her route, and they may make and contribute their own independent observations if they wish. "We get a lot of families participating," Larson says, and more than 75 percent of the state's routes are surveyed by groups of at least two.

Ỷ AGE RANGE All ages   **▦ TIME FRAME** Spring and summer
⊕ REGION North America   **$ FEE** None

Southwest Washington Amphibian Monitoring Project

www.swampproject.org

Based in Clark County, Washington, this project monitors area wetlands by examining the amphibian eggs found there. It's an independent project started by resident Peter Ritson in 2008, after seeing both the county's rapid urbanization and the positive conservation steps such as strong water-quality regulations. "There aren't enough biologists and researchers out there to collect the data to make informed decisions," Ritson says. "Lots of money is spent protecting water, and I wanted to see how effective those efforts are."

Photo by Peter Ritson

Southwest Washington Amphibian Monitoring Project participant Lisa Harlan in the field at La Center.

Ritson provides training in how to search for, identify, and count amphibian eggs, as well as the mechanics of the survey and how to report data. But participants make their own decisions on what sites to survey.

The grassroots, community-based nature of the project affords it possibilities that many other projects wouldn't have. "When we have citizens doing the work, it lets us get richer data in many cases," Ritson says. In a few instances, for example, landowners who wouldn't have let the state on their property have welcomed their neighbors to conduct the survey.

No previous knowledge or experience is required to participate; necessary training is provided. There are only four amphibian species in the area, so finding and identifying egg masses is relatively simple. "A person who is careful and interested can get a good count," Ritson says.

Surveys take place in late January through March, and participants are expected to conduct two four-hour pond surveys per winter. All ages are welcome, although Ritson recommends smaller sites for families with young children, who make good surveyors but may lack the focus and endurance for large wetlands.

For participants, the project offers the opportunity to experience the local environment in a way they probably haven't before. "In neighborhood parks, we're slogging through places where no one else goes," Ritson says. It also helps participants understand the local natural assets and the challenges they face—and to contribute important information to help improve the decisions about these natural places.

"There are big challenges, but it's very helpful to see where things are working and where they aren't," Ritson says.

AGE RANGE All ages **TIME FRAME** January through March **REGION** Southwest Washington **FEE** None

---

SNAKES

Carolina Herp Atlas   www.carolinaherpatlas.org

A project of the Davidson College Herpetology Laboratory, the Carolina Herp Access seeks to provide detailed information on the distribution of reptiles and amphibians of North and South Carolina.

"Knowing the geographic distribution of various species is very important in making land management decisions," says Michael Dorcas, director of Davidson's Herpetology Laboratory. There are big gaps in our understanding of species distributions, he adds, that citizen observations can help to fill.

After registering with the Atlas, participants simply report their observations, including location, species, date, and photos when possible. The atlas contains nearly twenty thousand records from more than one thousand users.

While land-use managers use the data the atlas has collected, it's also valuable for researchers in a variety of projects. For example, Davidson students recently conducted a study on urbanization and rattlesnake records, using only data that came from the atlas.

Contributing to the atlas requires minimal special knowledge, Dorcas says. While participants are asked to identify the species they submit, the Atlas's database manager vets the observations to correct misidentified photos or double-check records of species made outside of their known ranges or other observations that seem suspect. Thanks in part to these checks, the Atlas is a project that people at all levels can contribute to, whether they are serious amateur herpetologists or interested in nature more generally.

"Participants become more involved in the scientific process and more aware of what's important," Dorcas says. "It's an opportunity to teach kids and adults about local amphibians and reptiles."

♈ AGE RANGE All ages **▦ TIME FRAME** Year-round **⊕ REGION** North and South Carolina, however, the project may expand in the near future **$ FEE** None

Center for Snake Conservation
Snake Count

www.snakecount.org

Since 2011, the Center for Snake Conservation has sponsored annual snake counts in the fall to track snake distribution across North America. The counts last for about a week; participants can go anywhere they choose in that period and report the snakes that they encounter.

"Snakes play vital roles as mid- to top-level predators in our natural ecosystems, but they are often misunderstood and feared by humans," says Cameron

Young, founder and executive director of the Center for Snake Conservation.

Data collected is used to confirm the existence of rare species and to provide baseline data to monitor more common ones. The 2011 count recorded more than five hundred snakes in ninety-two species. Several rare species were among those photographed, including a black pine snake, a mole king snake, and a green water snake. Several other species were found in counties where they had never been observed before.

CSC is partnering with Project NOAH to help document snakes. (See Project NOAH, page 139.)

AGE RANGE All ages   **TIME FRAME** September   **REGION** North America   **FEE** None

---

Manitoba Herps Atlas

www.naturenorth.com/Herps/Manitoba_Herps_Atlas.html

Started in 2011, the Manitoba Herps Atlas is an effort to find out where reptiles and amphibians live in the province. "We need to build a better current database of species in Manitoba," says Doug Collicutt, who operates *Nature North*, an online nature magazine for Manitoba that hosts the atlas. "Most of the information that's out there is pretty old." The best reference on the province, *The Amphibians and Reptiles of Manitoba* by William Preston, was published in 1982, long enough ago that populations and their ranges may well have shifted.

Updating this information will help to make decisions about conservation and natural resources management. "We're trying to make the information public so the public can have the information needed for conservation decision-making," Collicutt says.

Only twenty-four species of reptiles and amphibians live in Manitoba, so minimal training is needed to identify them. Collicutt says the photos on the project's website should be sufficient in most cases, and the submitted data is vetted for accuracy before its included in the atlas. But the information is added fairly quickly. "To actually see your data points on a map is way cooler than waiting for a government report," Collicutt observes.

Beyond the type and location of the observation, participants report basic data about whether conditions, habitat type, and the life stages of the animals they see. Participants can also include photos or audio recordings of calls. Because it's a relatively simple process, it's suitable for all levels, from children to serious herp enthusiasts. Collicutt says many submissions are the result of observations made on family outings.

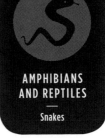

AMPHIBIANS
AND REPTILES
—
Snakes

AGE RANGE All ages **TIME FRAME** Primarily April through August
REGION Manitoba **FEE** None

Pennsylvania Herp Education and Resource Program

www.paherp.org/db

PA HERP maintains a citizen science database of the seventy-six reptiles and amphibian species in the state. "Most of these species are sparsely distributed over large areas, making it difficult to accurately survey for their presence and abundance," the project's website says. "However, large numbers of people involved in a variety of outdoor activities regularly come across individuals of these species."

Participants contribute their observations, as well as photographs or sound recordings of frog calls. The database currently has three thousand records.

"Data submitted on all tracked reptile and amphibian species will be organized by the Pennsylvania Herpetological Atlas and forwarded to the Pennsylvania Fish and Boat Commission and Pennsylvania Natural Heritage Program for entry into the PNHP database," the project's website says. "Thus, critical information on the location of these species will be available for conservation planning and decision making."

The project's website includes a section of kid's activities, as well as lesson plans for teachers.

AGE RANGE All ages **TIME FRAME** Spring and summer **REGION** North America
FEE None

Vermont Reptile and Amphibian Atlas

www.vtherpatlas.org

The Vermont Reptile and Amphibian Atlas has more than eighty-two thousand records, but it still has several goals. Current project priorities include perfecting distribution maps and monitoring for the potential discovery in the state of box turtles, Blanding's turtles, and marbled salamanders. The Atlas is also particularly interested in several rare species, including the Fowler's toad, North American racer, and Boreal chorus frog that have not been reported in recent years.

The atlas was originally started by the Reptile and Amphibian Scientific Advisory Group to the Vermont Endangered Species Committee, but it has become largely independent with the help of an organization called Vermont Family Forests. It shares information with colleges and museums, as well as the state Nongame and Natural Heritage Program for conservation purposes.

Anyone can contribute observations from anywhere in the state. Observations include a photo or description of the animal, details about the location and habitat of the spotting, and the date.

The atlas also serves as an opportunity to nurture appreciation of reptiles and amphibians. "In order to care about anything, you have to be introduced first," says Jim Andrews, coordinator of the atlas. Many of the observations in the atlas, he adds, come from people who spot something they haven't seen before and submit their observation in hopes of learning more.

Knowing what animals live where is critical for land managers to be able to make land-use decisions while still protecting species of concern. Improved information can save money, because it allows for conservation action to be taken earlier. Citizen observers often become advocates, Andrews says. "People develop a sense of ownership and stewardship of species they report. If these species are to survive they need local stewards."

Contributing to the atlas is an activity suitable for all ages. "Many parents see it as a vehicle to get their kids out and do some natural history," Andrews says. Many children are drawn to this kind of exploration, and one presentation Andrews made to an elementary school class ultimately led to a student making a career in reptile research. The project's website includes a section aimed at teachers on how to use the project as an educational tool.

AGE RANGE All ages   **TIME FRAME** Year-round   **REGION** Vermont   **FEE** None

Hawaii Sea Turtle Monitoring

www.inaturalist.org/projects/hawaii-sea-turtle-monitoring

AMPHIBIANS AND REPTILES

Turtles

Presented by the National Oceanic and Atmospheric Administration (NOAA)and hosted on iNaturalist, the Hawaii Sea Turtle Monitoring project monitors green and hawksbill turtles and invasive algae in Hawaii. Of particular interest to the project is observations of turtles with tumors, a significant cause of mortality among green turtles.

"There's a lot of casual enthusiasm about sea turtles," says Kyle Van Houtan, head of the Marine Turtle Assessment Program for the Pacific Islands region and the project's administrator. "Tourists love to see and photograph them—they're charismatic creatures."

But hawksbills in Hawaii, in particular, are rare: Van Houtan says there have never been records of more than twenty individuals nesting in a single year. Therefore, "Any information we get is interesting for science and relevant for management." In the past decade, two hawksbills had been stranded alive in Hawaii; a third record came through iNaturalist and the monitoring project. Van Houtan has used data from the project in a 2012 research paper.

NOAA hopes to lead scientific research and management efforts in other countries, and Van Houtan says that iNaturalist can help. Many other nations don't have highly developed infrastructures, but residents often have cell phones with cameras, which is all the equipment needed to contribute.

Like other projects on iNaturalist, participation is open to anyone with a free account, although observations must be made in Hawaii. Participants post photos of their observations with information about its time and location.

AGE RANGE All ages **TIME FRAME** Year-round **REGION** Hawaii **FEE** None

---

Neighborhood Box Turtle Watch www.naturalsciences.org/research
-collections/citizen-science/neighborhood-box-turtle-watch

The North Carolina Museum of Natural Sciences' Neighborhood Box Turtle Watch is a database of wild box turtles, particularly those living in suburban areas, started in 2008. "It is important to collect information about box turtle populations living in proximity to residential development to determine the viability and health of these populations," says Jerry Reynolds, the museum's senior manager of outreach. In addition to documenting box turtles in suburban areas, other goals of the project include improving appreciation of turtles and making neighborhoods more turtle-friendly.

Participants simply need to report box turtles they find, along with photos of the top and bottom of the shell and notes about the turtle's physical attributes and activity. Photographs are used to identify individual turtles so they can be recognized if they are observed again. Information submitted to the museum is maintained by the museum's research section with the potential for future publication. Reynolds also encourages participants to submit their observations to the Carolina Herp Atlas to help map box turtle distribution.

Reynolds says that the project is excellent for all ages, with kids being able to search for turtles and document their finds. "Exploring nature is a great way to develop a better awareness and appreciation for the wonder of nature," Reynolds says. "Participating in research gives kids an idea of how we gather data and interpret results from that data that may give us some new information on what we are researching."

AGE RANGE All ages   **TIME FRAME** Spring through fall   **REGION** North Carolina   **FEE** None

Resources

ADULT BOOKS

A Field Guide to Western Reptiles and Amphibians by Robert Stebbins and Roger Tory Peterson (Houghton Mifflin Harcourt, 2003).

Frogs: A Chorus of Colors by John and Deborah Behler (Sterling, 2008).

The Frogs and Toads of North America by Lang Elliott, Carl Gerhardt, and Carlos Davidson (Mariner, 2009).

A Natural History of Amphibians by Robert Stebbins and Nathan Cohen (Princeton University Press, 1997).

Snakes of North America: Eastern and Central Regions (Lone Star Field Guides) by Alan Tennant (Taylor Trade Publishing, 2003).

Snakes of the United States and Canada by Carl and Evelyn Ernst (Smithsonian, 2003).

Tadpoles: The Biology of Anuran Larvae, edited by Roy
McDiarmid and Ronald Altig (University of Chicago
Press, 2000).

BIRDS
—

CHILDREN'S BOOKS

The Case of the Vanishing Golden Frogs: A Scientific Mystery
by Sandra Markle (Millbrook, 2011). Ages 10–14.

Frogs! Strange and Wonderful by Laurence Pringle, illustrated by Meryl Henderson
(Boyds Mills, 2012). Ages 6–10.

Scholastic True or False: Amphibians by Melvin and Gilda Berger (Scholastic, 2011).
Ages 7–9.

Scholastic True or False: Reptiles by Melvin and Gilda Berger (Scholastic, 2008). Ages
7–9.

Sea Turtles by Laura Marsh (National Geographic, 2011). Ages 7–10.

The Snake Scientist by Sy Montgomery (Houghton Mifflin, 2001). Ages 10–15.

Snakes by Gail Gibbons (Holiday, 2008). Ages 6–8.

Turtle, Turtle, Watch Out! by April Pulley Sayre. Chalesbridge, 2010). Ages 5–9.

BIRDS

Birds have an extremely long history as the subject of citizen science research,
and with good reason: the skills needed for amateur bird-watching are highly
transferrable to the collecting of other types of scientific data. This section
includes projects collecting information about all types of birds as well as
those focused on a several specific species, including cranes, loons and hawks.
If you are particularly interested in shorebirds, see "Beaches, Wetlands, and
Waterways," page 117.

Annual Midwest Crane Count

www.savingcranes.org/annual-midwest-crane-count.html

Sponsored by the International Crane Foundation, the Annual Midwest
Crane Count involves more than three thousand participants in Wisconsin,
Illinois, Indiana, Iowa, Michigan, and Minnesota.

"Sandhill cranes once experienced severe population declines in the late
1800s to early 1900s in the upper Midwest, but have recovered successfully,"

according to the Foundation. "The Annual Midwest Crane Count has documented the growth of the sandhill crane population, and allows ICF to monitor crane abundance and distribution."

Participants count the sandhill cranes they see or hear at a given site on a single day in April. They also report any breeding activity they see, and any banded sandhill or whooping cranes.

To get involved, see the project's website to determine if your county is involved. If not, you can volunteer to become a county coordinator and bring the count to your area.

AGE RANGE All ages   **TIME FRAME** April   **REGION** Upper Midwest   **$ FEE** None

Audubon Society Christmas Bird Count

http://birds.audubon.org/christmas-bird-count

The Audubon Society's annual Christmas Bird Count (CBC) is the world's oldest citizen science project. It dates back to 1900, when ornithologist Frank Chapman proposed counting birds as an alternative to the then-traditional Christmas bird hunts. It takes place in late December and early January when participants follow set routes through a designated fifteen-mile circle, counting every bird they see or hear. There are more than two thousand count circles, and each circle has several counters to cover more ground and to ensure that new counters can be teamed with experienced experts.

Data from the Christmas Bird Count provides a wealth of information about the long-term health of bird populations—which can serve as an indicator of broader environmental threats. The data has been put into use in several instances. In the 1980s, the Christmas Bird Count helped to quantify the decline in numbers of American black ducks due to overhunting and habitat loss. CBC data was used to justify hunting regulations that helped to stabilize populations. "CBC is the backbone of scores of peer reviewed studies, state and local conservation plans, and headline-making news," wrote Audubon president and CEO David Yarnold in his 2011 letter to count participants. "When Audubon's 2009 Birds & Climate Change Report revealed that many species in North America shifted their ranges north on an average of one mile a year because of warming temperatures, Katie Couric featured it on the *CBS Evening News*. CBC inspired Congress to pass vital legislation such as the Migratory Bird Treaty Act that benefits species across the Western hemisphere." In addition to identifying species at risk, the CBC has helped to document conservation successes, such as the bald eagle and the peregrine falcon.

To participate, check for available Christmas Bird Count circles at the "Get Involved" page on the Christmas Bird Count website beginning in late fall.

AGE RANGE All ages   **TIME FRAME** December 14 through January 5
REGION US and Canada, with some additional sites in the Western Hemisphere
FEE $5 per participant; fee is waived for children under 18

Bird Conservation Network Survey

www.bcnbirds.org/census.html

BCN has surveyed birds in the Chicago area, including sites in southern Wisconsin and northwestern Indiana, since 1998. "One aim is to compile a record of bird populations, particularly in breeding season, throughout the Chicagoland area," says Lee Ramsey, who manages the survey with Judy Pollock of the Chicago Audubon Society. The survey also provides information to the stewards who are managing the many natural restorations in the area. "My job is to warn about possible impacts, both good and bad," Ramsey says.

BCN works with participants to assign convenient but important sites for monitoring. The survey consists of five-minute counts of all birds heard or seen within seventy-five meters of several count points on the site. Two survey visits at sunrise during June, the prime breeding season for birds locally, are required, and BCN asks participants to make at least five visits throughout year to monitor migration and wintering populations.

Strong field identification skills are required. Ramsey says participants particularly need to be able to identify bird songs and calls, since many birds will be hard to see during breeding season. Ramsey also seeks participants willing to commit to the project over several years. "What we want ideally is continual reports over as many years as possible from the same person," he says.

All data is also shared with eBird (see page 69) to allow access for a wide range of research purposes.

BIRDS

BCN has also sponsored a number of short-term blitzes, in which volunteers collected data about specific habitats or individual sites. In 2012, the blitz focused on red-headed woodpeckers, a species of particular conservation concern.

🏃 **AGE RANGE** Primarily adults 📅 **TIME FRAME** Year-round, with special focus in June 🌐 **REGION** Chicago and its suburbs 💲 **FEE** None

Breeding Landbird Monitoring Project

See Vermont Forest Bird Monitoring Program, page 81.

Celebrate Urban Birds

http://celebrateurbanbirds.org

Celebrate Urban Birds is a project from the Cornell Lab of Ornithology in which participants make ten-minute observations of the birds in their neighborhood. The project focuses on sixteen species that are easy to identify and that can be found in urban, suburban, or rural areas, with the intention that it be easy for beginners to take part.

"When kids participate, they're thrilled to be doing something that's real," says Karen Purcell, coordinator of the program. "They suddenly have an excuse to be outside and to look at the natural world in a careful way." Those careful examinations are often difficult to make time for. Many people, Purcell says, are so accustomed to being constantly on the go that simply being still for ten minutes is an incredible challenge.

Research goals of the project include investigating how birds use green spaces and how the size and quality of the sites affect them, in both urban and suburban environments. Ultimately, Purcell adds, the project will build enough of a collection of data to evaluate how populations have changed over time.

There's a strong educational component to the project as well. "What educators tell us is that as soon as people start to participate, they start to ask questions," Purcell says. Those questions can spark a lifelong interest in birds and their protection. "Once people start getting hooked, then they care, and 'conservation' means something because they have a connection" to the birds, she adds.

BIRDS

Celebrate Urban Birds works with many types of organizations, including businesses, health care organizations, schools, libraries, and senior centers. The program offers mini-grants to these organizations annually, encouraging efforts to use the program to help educate the public about birds, improve or create bird habitats through large or small planting projects, and conduct outreach through the arts. "If we send one of our scientists to Central Park to look at birds with binoculars and a data form, he's not going to attract a crowd," Purcell says. "If you ask a community artist to create a giant mural or cardboard birds to parade through the park, that does attract folks."

AGE RANGE All ages   **TIME FRAME** Year-round   **REGION** North America
FEE None

---

eBird
http://ebird.org

Launched in 2002 by the Cornell Lab of Ornithology and the National Audubon Society, eBird is a massive online database of bird observations. "It is amassing one of the largest and fastest growing biodiversity data resources in existence," the project's website says, with more than 3.1 million observations added in March 2012 alone.

"The study of birds is one of the fields of science where amateurs can have the biggest impact," says Marshall Iliff, eBird project leader. Anyone, professional or amateur, is welcome to contribute to eBird. The site is suitable for all ages, he adds. "Some of our best users are actually pretty young kids," Iliff says.

Participants must register online, although the registration is good for any of Cornell's citizen science projects. They report basic data about the birds they see, including the date, location, type and number of birds seen, and length of time spent observing.

All this information is accessible to anyone who is interested. Since the quantity of data is so vast, it is extremely useful to scientists answering questions about the distribution and quantities of specific species of birds. "Every time that you see and identify a bird, you are holding a piece of a puzzle," the eBird website says. "Whether you are casually watching birds in your

BIRDS

backyard, or chasing rare species across the country, you are helping to put this puzzle together." The data has also been used to identify precise migratory paths, track how the distribution of a species changes, or monitor breeding and wintering ranges. Dozens of scientific papers have cited eBird data.

For beginning birders seeking to gain knowledge, "one of the best things you can do is keep detailed records," Iliff says. Participants can use the site for their own tracking, maintaining life lists of the birds they have seen. It also allows mapping bird information on a small scale, so users can get lists of birds that they should see by county, and even by specific site. If a user reports something that is out of its expected range, a local expert can investigate to determine if the user made an error or found something genuinely unusual. In that way, the site helps to connect birders. "It puts the newest birders in touch with experts in their region," Iliff says.

There are a number of specialized eBird portals that contribute to the database. Some are regionally focused, while others focus on specific birds or follow different observation methods to achieve different goals. Each of these portals is fully integrated into the eBird database, so their information can be compared with the full set of eBird observations or separately. See the full list of portals at ebird.org/content/ebird/about/ebird-regional-portals.

♈ AGE RANGE All ages   **▦ TIME FRAME** Year-round   **⊕ REGION** Worldwide   **$ FEE** None

---

Great Backyard Bird Count   www.birdsource.org/gbbc

The Great Backyard Bird Count is an annual four-day bird count in February led by the Cornell Lab of Ornithology, the Audubon Society, and Bird Studies Canada. Participants can count birds at any location for as long as they wish during the count to help create a real-time snapshot of where birds are throughout the continent. "Bird populations are dynamic; they are constantly in flux. No single scientist or team of scientists could hope to document the complex distribution and movements of so many species in such a short time," the project's website says.

Participants report the largest number of birds of each species they see together at any one time. They also give details about the location and habitat, weather conditions, and the time spent counting. Data is contributed to the Avian Knowledge Network database.

In 2012, the GBBC received more than one hundred thousand reports with more than 17 million bird observations. Among the unusual observations were a mile-long flock of tree swallows in Ruskin, Florida, record numbers of snowy owls that likely came south due to low populations of prey in the Arctic, and significant numbers of waterbirds spending the winter farther north than usual as the mild winter left ponds and lakes unfrozen.

The GBBC welcomes observations from all ages, and the project website has a section to guide kids in participating. The project also welcomes groups to make and share observations together, and many local groups sponsor birdwalks as part of the GBBC.

AGE RANGE All ages   **TIME FRAME** February   **REGION** US and Canada   **$ FEE** None

---

Hawk Migration Association of North America

www.hmana.org

HMANA coordinates a network of hawk watches—monitoring sites where raptors are known to aggregate during migrations, and where observers collect data to determine long-term population trends.

"Raptors can be quite elusive and it's difficult to monitor their populations with surveys like breeding bird surveys and Christmas bird counts," says Julie Brown, monitoring site coordinator. "Monitoring raptors during fall and spring migration is the best opportunity to collect information on their patterns and over time, these observations can be used to assess populations."

Individual hawk watches are independent and determine their own policies and procedures, although HMANA provides a formal standard protocol that can be customized for each site to ensure observations are useful as scientific data. Recommendations include frequent and regular observations—ideally daily—at a standard time and location.

Data collected is included in the HawkCount.org database, the Avian Knowledge Network, and the Global Biodiversity Information Facility. Hawk Count.org is a critical part of the Raptor Population Index program, which

BIRDS

assesses and publicizes raptor population trends. While the RPI shows that most raptor populations—including previously endangered species such as the bald eagle and peregrine falcon—have grown in population over the last three decades, American kestrels have shown sharp declines.

Participation does require some identification skills, although in most cases new participants can learn from experienced hawk watchers. Families are also generally welcome. "Children have the best eyes for picking raptors out of the sky, and they love the challenge," Brown says.

AGE RANGE All ages   **TIME FRAME** Fall and spring   **REGION** North America
FEE None

Minnesota Loon Monitoring Program

www.dnr.state.mn.us/eco/nongame/projects/mlmp_state.html

Started in 1994, the Minnesota Loon Monitoring Program (MLMP) is a monitoring program to detect stability in the population of the common loon. "Not only is it our state bird, but Minnesota has the largest breeding loon population in the lower forty-eight states," says Krista Larson, research biologist with the Minnesota Department of Natural Resources and coordinator of the program. "We have a responsibility to maintain healthy populations of this species within our borders."

Hundreds of volunteer participants survey a lake or multiple lakes one time per year in early July. About six hundred lakes are surveyed in total. Participants report the number of loons they observe, as well as data about weather and shoreline conditions. That information gives up-to-date information about any changes in loon populations and insight into the causes of any declines that might occur.

MLMP data have other value as well. "One unexpected way it is being used is as a baseline to assess whether loon populations in Minnesota were impacted from the Gulf Coast oil spill," Larson says. Most loons from the state winter along the Gulf Coast, and subadult loons do not return to Minnesota until their third year. Loons that hatched in Minnesota in 2008 or 2009 would have been in the Gulf during the 2010 Deepwater Horizon spill, and information about how populations have fared is now being gathered.

Program data is also a good indicator of overall lake health. "Loons thrive in clear lakes that have healthy fish populations and undisturbed shorelines with plenty of natural vegetation, so loons are a good barometer to measure the overall quality of Minnesota's lakes," Larson says. She adds that protecting the natural environment and caring for those lakes is a motivating factor for many participants to be a part of the program.

The program has happy news to report: the loon population has been stable in the state since the program's inception. Regular monitoring is important to help ensure that state of affairs, however. "If you don't have baseline information, you can't detect declines," Larson says.

Photo by Roland Jordahl

Roland Jordahl monitoring loons at Anderson Lake for the Minnesota Loon Monitoring Program.

BIRDS

No formal training is required to contribute to the MLMP; participants simply need to be able to access lakes by boat, canoe, kayak, or foot, and count loons. Larson says that makes it easily accessible for families. "Many of our volunteers are retirees or citizens who live on a particular lake that is part of the monitoring program," Larson says. "Many of them bring along spouses and children to help with the counts."

Y AGE RANGE All ages   **TIME FRAME** July   ⊕ **REGION** Minnesota   **$ FEE** None

Mountain Birdwatch   www.vtecostudies.org/MBW

Launched in 2000, Mountain Birdwatch is a long-term monitoring program for high-elevation forest birds. Because the areas covered are so remote, major bird surveys did not cover them, says Judith Scarl, the project coordinator. It began with a focus on Bicknell's thrush, the only songbird endemic to northeastern mountain forests and a vulnerable species due to its restricted range. Before 2000, the surveys of Bicknell's thrush weren't coordinated or standardized, and some areas were missed, Scarl says.

Operated by the Vermont Center for Ecostudies, the project has expanded to include sites in New York, New Hampshire, Maine, Quebec, and Canada's Maritime Provinces. As of 2010, MBW tracks ten songbird species, as well as red squirrel populations.

Climate change is a significant threat for the mountain forests surveyed by the project and the animals that live there. As temperatures warm, most animals can adapt by moving north or to higher elevations. "But when you're on top of a mountain, you have nowhere to go," Scarl observes.

Participants are assigned one of 130 routes to survey. Each citizen scientist surveys his or her route one time during June in the early mornings. Each route has three to six set points where participants count all target species for four five-minute periods. While small groups or families are welcome to walk the route, only a single person may perform the actual bird count.

While there is no age limit, Mountain Birdwatch is a fairly demanding project that is best suited for older kids. While there is variety in the routes, many involve serious hiking, and the data collection protocols may be a challenge for younger volunteers. But it also offers unique rewards. On some routes,

BIRDS

Scarl says, "The volunteer may be the first person to walk through an area that year since the snow melted." Many participants will conduct their surveys as part of a weekend camping trip.

Mountain Birdwatch holds annual training workshops to teach volunteers about biology and the unique high-elevation environment, as well as the scientific reasoning behind the project. Each participant also receives a training manual and a CD of the various species' songs and calls.

The data generated by the project is available online at the Avian Knowledge Network. It has been used by nonprofits, government groups, and landowners to make conservation and land-use decisions, such as in siting wind farms or expanding ski areas.

AGE RANGE All ages, but better suited for older kids and up **TIME FRAME** June **REGION** Vermont, New York, New Hampshire, Maine, Quebec, and Canada's Maritime Provinces **$ FEE** None

Neighborhood Nestwatch

http://nationalzoo.si.edu/SCBI/MigratoryBirds/
Research/Neighborhood_Nestwatch

Neighborhood Nestwatch is a citizen science project operated by the Smithsonian Migratory Bird Center to observe the birds that live among humans. Participants find and monitor bird nests, particularly in their own backyards, to collect information about the success of these nests in nurturing live young and how long backyard birds live.

More than two hundred citizen scientists currently participate in Neighborhood Nestwatch. Participants observe nests several times during the nesting cycle, collecting data on the type of bird, where it built its nest, the number of eggs laid, and whether fledglings hatched. There is a waiting list to join, because Nestwatch staff visits each participant's property to band birds and provide monitoring training, although the project also accepts data from "at-large" participants who monitor nests without formal training. Participants include people of all ages, including families with young children, and all levels of birdwatching experience.

BIRDS
—

More than seven thousand birds have been banded through the project. Data from Neighborhood Nestwatch has been used in a number of research projects on topics ranging from West Nile Virus to lead levels in birds to the types of nest predators in rural and urban areas.

AGE RANGE All ages   **TIME FRAME** Primarily spring and early summer
REGION Within sixty miles of Washington, D.C.   **FEE** None

NestWatch   www.nestwatch.org

NestWatch is a nationwide program to monitor bird nesting sponsored by the Cornell Lab of Ornithology. "Without the help of NestWatch volunteers, it would be impossible to gather enough information to accurately detect and track changes in the reproductive biology of birds at the large scales of states, regions, or the entire US," says Jason Martin, NestWatch project leader. "Specifically, the NestWatch data is appropriate to detect and monitor potential changes in timing and success of breeding that coincide with climate change."

After taking a short quiz online to learn how to monitor nests without harming them or the birds in them, participants find active bird nests, primarily of about forty focal species, and visit them every three to four days. They collect data about the number of eggs and young, the nest's location, status of the nest and its inhabitants, and the activity of any cowbirds that may lay their eggs in other birds' nests.

The NestWatch website tracks each participant's observations for personal usage, while also adding them to a database that has been cited in numerous research papers. More than 411,000 observations from more than 142,000 nests are included in the database.

"While contributing extremely valuable information to science, NestWatch volunteers also learn firsthand about birds and form a lifelong connection with the natural world," Martin says. "Not only do they make an incredibly valuable contribution to science, but they also directly witness and learn about the often unseen family life of birds, from eggs, to fuzzy chicks, to gawky youngsters ready to take their first fluttering flight." Martin adds that the project is suitable for all ages, although young children should be accompanied by an adult.

NestWatch has a number of chapters based at nature-oriented organizations throughout the United States. Chapters help to recruit and train NestWatch participants and conduct monitoring as part of their own missions.

⌀ AGE RANGE All ages **▦ TIME FRAME** February through October **⊕ REGION** North America **$ FEE** None

North American Breeding Bird Survey   www.pwrc.usgs.gov/bbs

Coordinated by the Geological Survey, the Canadian Wildlife Service, and since 2007, the Mexican National Commission for the Knowledge and Use of Biodiversity, the North American Breeding Bird Survey was started in 1966 to investigate potential effects of increased pesticide spraying on continental bird populations. Since then, the project's scope has expanded to include most landscape-level threats, with the BBS now tracking and reporting annual population changes of more than 420 North American bird species at a variety of geographic scales.

"A main selling point of the BBS is that each participant is directly contributing to national avian management and conservation efforts through local action," says Keith Pardieck, US coordinator of the survey. Local bird counts are aggregated to provide regional population information that, along with other indicators, are used to assess and inform national avian management priorities. In addition, many of the species tracked in this survey aren't monitored by any other means.

The BBS is a roadside survey with a statistically rigorous, standardized design. Randomly established routes are sampled by more than 2,200 citizen scientists during the bird breeding season, primarily in June. Routes include fifty stops, roughly half a mile apart. At each stop, surveyors identify and count each bird seen or heard within a quarter mile in three minutes. Participants also count the motorized vehicles that pass each stop during that time. As with many survey projects, only one person may identify and count birds, although assistants may accompany the counter to assist with recording, car counting, driving, or other tasks.

"People get attached to their routes," Pardieck says. "They adopt the route and look forward to going every year" and bearing witness to the changes taking place over time.

BIRDS

Participating in the survey requires advanced bird identification skills, particularly by sound, which typically takes years to develop. Surveys also begin before dawn, so the survey is not a natural fit for families with young children interested in a casual citizen science activity. Pardieck says there are some kids, primarily teens, who do participate, most often by assisting a parent or mentor who is already a skilled birder. "If kids have a keen interest in birds, they should find the BBS an enjoyable experience when shared with an enthusiastic and knowledgeable adult," he adds.

The BBS provides avian population trend estimates, relative abundance estimates, and summer distribution data that are used by the US Fish and Wildlife Service, Canadian Wildlife Service, state wildlife agencies, and many other organizations to identify conservation priorities and research or management needs. In addition, more than four hundred scientific publications have relied on BBS data.

AGE RANGE Teens and up   **TIME FRAME** June   **REGION** Continental US, Alaska, southern Canada, and northern Mexico **$ FEE** None

---

Operation RubyThroat: The Hummingbird Project

www.rubythroat.org

Ruby-throated hummingbirds are the most common hummingbird in the world. "Even though they're really common, we don't know much about them," says Bill Hilton Jr., the project's principal investigator. Science knows almost nothing about their migration, and no doctorate has ever been written about their breeding behavior because their nests are so difficult to find. "Citizen science, with a lot more eyes out there, might be able to find enough nests for a study," Hilton says.

Hilton started Operation RubyThroat in 2002 to help study the ruby-throated hummingbird. Participants in the United States, Canada, Mexico, and Central America participate by collecting data about the rubythroat's migration, arrival and departure dates, nesting, and feeding behavior. They contribute the data either through the GLOBE Program or directly through Operation RubyThroat itself.

Operation RubyThroat welcomes participation from all ages. It was originally designed for classroom use, although that proved less than ideal because the rubythroat spends much of the school year in Central America.

Somewhat connected to the program, Hilton has led annual citizen science research trips to Central America since 2004 to study the rubythroat on their winter grounds. These trips, suitable for adults only, include observations of rubythroats in the morning and other field trips in the afternoons. "Citizen scientists make a real impact in the world of science," he says. "We have fun, but we go out in the field every morning." Those trips have raised an interesting question by discovering that rubythroats are territorial on their wintering grounds. While the behavior makes sense on their springtime breeding grounds, it's not clear why they would demonstrate it when mating isn't occurring.

He has also demonstrated that rubythroats show site fidelity, returning to the same breeding and wintering grounds each year. "It gives us the argument to protect their habitat in both locations, because they're a shared resource," Hilton says.

AGE RANGE All ages **TIME FRAME** Spring through fall in the US and Canada; winter in Mexico and Central America **REGION** Eastern North America **FEE** $30 for school groups, $20 for homeschool groups

Project FeederWatch
www.birds.cornell.edu/pfw

Project FeederWatch is a winter survey of birds that visit feeders throughout North America to help track movements of winter bird populations. It grew out of the Ontario Bird Feeder Survey, which was founded in 1976, and is now operated by the Cornell Lab of Ornithology and Bird Studies Canada.

"FeederWatch is designed to look at changes in the distribution and abundance of birds over time," says David Bonter, project leader.

Participants choose their own site to count birds that appear because of the food or water they have provided. Each count period lasts two days, and participants count the largest number of individuals of each species seen at one time during that period. They also record weather conditions and the amount

of time they spend watching their feeders; during a count period, participants may observe their feeders as much as they choose.

Participants can schedule count periods up to weekly during the twenty-one-week FeederWatch season; the only restriction is that count periods must be at least five days apart.

FeederWatch data indicates range expansions of many of its target species. "Many species that visit feeders are well adapted to human landscapes, so they're doing well," Bonter says. An exception is the evening grosbeak, whose range has contracted and whose numbers have decreased by half in locations where it can still be found. The project also provides valuable information about house finches, whose populations were devastated by disease in the 1990s. Because the project began before the disease struck, Bonter says, "we know how populations were doing before, during, and after." About a dozen scientific papers have used the project's house finch data alone.

All data goes into a central database accessible to researchers and online. The project also provides information to participants in the form of graphs, maps, and articles.

The project's website includes extensive information designed to help teachers and homeschoolers incorporate the project into their curriculum. "We find participants often increase their knowledge of birds and nature as part of the project," Bonter says. Many other participants are grandparents, and Bonter says he regularly hears anecdotes from them about how they share the experience with their grandchildren.

AGE RANGE All ages   **TIME FRAME** November through April
REGION North America   **FEE** $15 (CAN$35 for Canadian participants), which includes a research kit and a subscription to *Living Bird News* or *BirdWatch Canada*.
Participants provide their own feeder and seed.

Seattle Audubon Society Neighborhood Bird Project
www.seattleaudubon.org

Started in 1994, the Neighborhood Bird Project monitors urban birds in King County, Washington, and advocates for urban wildlife habitats. Participants conduct monthly bird censuses at nine city parks. Each park is surveyed by teams that divide the site into loops with several counting stations each. At each counting station, team members count every bird species seen, heard, or flying over within fifty meters for five minutes.

Most count sites welcome any level of knowledge among participants and are suitable for families to participate in, even with young children. A couple do require advanced skills, however; see the project's website for details.

Advanced birders in Seattle can also participate in Seattle Audubon's Puget Sound Seabird Survey.

AGE RANGE All ages **TIME FRAME** Year-round **REGION** King County, Washington **FEE** None

---

Vermont Forest Bird Monitoring Program

www.vtecostudies.org/FBMP

http://science.nature.nps.gov/im/units/NETN/monitor/programs/
breedinglandbird/breedinglandbird.cfm

Started in 1989, the Vermont Forest Bird Monitoring Project seeks to track long-term changes in songbird populations. Currently twenty-nine sites, representing nine different habitats, are monitored each June, when the birds are breeding. Citizen scientists visit each site twice and record all the birds seen or heard.

"The FBMP is sort of an early-warning system for forest birds," says Steve Faccio, conservation biologist for the Vermont Center for Ecostudies. "By monitoring their relative abundance over the long term, we are able to detect population trends for each species that allows us to assess their status and, if needed, take action before they are critically endangered."

While many studies have observed population declines in many songbird species, they have focused primarily on fragmented landscapes. The FBMP is one of the few studies to collect data from protected forests in the Northeast where large tracts of undisturbed forest habitats remain.

Faccio cites the Canada warbler as a success story from the project: population declines noted over ten years of monitoring led to some in-depth research projects. Those projects improved understanding of the species' habitat requirements and a guide to forestry practices that can maintain it.

BIRDS

The FBMP is less suited to young children than many projects. Participants require excellent skills in bird identification by both sight and sound. Surveys require two to four hours in the early morning, plus another hour to record data. Participants are expected to commit to the project for multiple years. But for skilled birders, Faccio says, the project offers an opportunity to work for conservation that will benefit the birds they are passionate about.

Faccio coordinates two distinct, but similar, forest bird monitoring projects. In addition to the Vermont survey, the Breeding Landbird Monitoring Project surveys birds at eleven national parks in the Northeast from New Jersey to Maine.

AGE RANGE Primarily adults   **TIME FRAME** June   **REGION** Vermont and northeastern US   **$ FEE** None

---

Vermont Loon Recovery Project   www.vtecostudies.org/loons

In the mid-1980s, the common loon was not so common in Vermont. Due to disturbance of nesting sites, loss of habitat, and contamination by heavy metals, there were only twenty-nine adults counted in the state, and only seven breeding pairs.

The Vermont Loon Recovery Project was started by the Vermont Center for Ecostudies in 1978 to reverse the loon's decline, with significant success. The loon was designated as an endangered species in Vermont in 1987, but removed in 2005 as the population grew. During a one-day census in July 2011, citizen scientists counted 271 adult loons—approaching the possible maximum number of this extremely territorial bird the state's lakes can support.

Still, human impact could quickly threaten the loon's population again. "Here in Vermont, half of our loons are still nesting in sites that are at risk," says Eric Hanson, the project's coordinator. "We're always going to have some management need."

The project's citizen science efforts are working to monitor loon populations to ensure they remain healthy. There are a few levels to these efforts. LoonWatch is the one-day census, taking place the third Saturday in July. Participants are assigned one of more than 150 lakes to monitor for one hour

BIRDS

to count loon populations, as well as the presence of chicks or subadult loons, and any ospreys or bald eagles sighted. The project also accepts casual observations of loons, particularly at lakes that have no regular volunteer monitoring. These surveys can take place at any time between May and October.

The most in-depth citizen science opportunity is the project's Adopt-a-Lake program. Participants are assigned a lake with nesting or territorial pairs to monitor one to four times per month from mid-May through August. Volunteers make observations about nesting and hatching activity, as well as predators or human disturbances.

All data is contributed to Vermont's eBird portal (see eBird, page 69) for broader research as well as the state's conservation efforts.

Monitoring gives current information about where breeding pairs may be forming. When they do form, "I can be proactive in working with landowners" to protect them, Hanson says. Project volunteers also help post warning signs near nests to alert the public and prevent inadvertent disturbances, and build and maintain nesting rafts for lakes where lakefront development doesn't leave adequate loon nesting habitat.

Some 250 people participate in the project at some level. Hanson says that the one-day census is particularly popular as a family activity, even becoming a regular tradition for many.

Hanson hopes to expand the project into a broader initiative for lake and shoreline health. "People have a strong connection to loons," he observes." You can use them to get people to appreciate the other things that need conservation."

AGE RANGE All ages **TIME FRAME** May through October **REGION** Vermont **FEE** None

---

YardMap

www.yardmap.org

YardMap is a project from the Cornell Lab of Ornithology that seeks to identify landscape management techniques that improve value for wildlife. Participants map their yards, or other locations, to show distinct habitats and other objects found there.

BIRDS
—

The project's aims are heavily weighted toward advocacy. "Since most of the landscaping in the average yard (lawn and impervious area) are relatively useless for birds, we think it's worth starting a conversation about how to change practices in the average yard to put that space to use for birds," the project's website says.

Ⓨ AGE RANGE All ages   **▦ TIME FRAME** Year-round   **⊕ REGION** Worldwide   **$ FEE** None

Resources

ADULT BOOKS

The Audubon Society Guide to Attracting Birds: Creating Natural Habitats for Properties Large and Small by Stephen Kress (Comstock, 2006).

Bird Sense: What It's Like to Be a Bird by Tim Birkhead (Walker, 2012).

The Bluebird Effect: Uncommon Bonds with Common Birds by Julie Zickefoose (Houghton Mifflin Harcourt, 2012).

Call of the Loon by David Evers (Willow Creek Press, 2006).

Cerulean Blues: A Personal Search for a Vanishing Songbird by Katie Fallon (Ruka Press, 2011).

The Common Loon: Spirit of the Northern Lakes by Judith McIntyre (University of Minnesota Press, 1988).

Falcons of North America by Kate Davis (Mountain Press, 2008).

A Field Guide to Hawks of North America by William Clark and Brian Wheeler (Houghton Mifflin Harcourt, 2001).

Hawks from Every Angle: How to Identify Raptors in Flight by Jerry Liguori (Princeton University Press, 2005).

How to Be a Better Birder by Derek Lovitch (Princeton University Press, 2012).

Loon Magic by Tom Klein (Paper Birch Press, 1985).

Migrating Raptors of the World: Their Ecology and Conservation by Keith Bildstein (Comstock Publishing, 2006).

Songbird Journeys: Four Seasons in the Lives of Migratory Birds by Miyoku Chu (Walker, 2007).

CHILDREN'S BOOKS

Backyard Birds by Karen Stray Nolting and Jonathan Latimer (Sandpiper, 1999). Ages 10–14.

Bird Talk: What Birds Are Saying and Why by Lita Judge (Flash Point, 2012). Ages 6–9.

Bird Songs by Betsy Franco (Margaret K. McElderry, 2007). Ages 5–7.

Bring on the Birds by Susan Stockdale (Peachtree, 2011). Ages 4–6.

Even an Ostrich Needs a Nest: Where Birds Begin by Irene Kelly (Holiday, 2009). Ages 5–9.

Hummingbirds by Melissa Gish (Creative Education, 2012). Ages 4–6.

Loon Magic for Kids by Tom Klein (Northword Press, 1991). Ages 6–8.

Moonbird: A Year in the Wind with the Great Survivor B95 (Farrar, 2012). Ages 6–10.

Olivia's Birds: Saving the Gulf by Olivia Bouler (Sterling, 2011). Ages 9–12

Pale Male: Citizen Hawk of New York City by Janet Schulman (Knopf, 2008). Ages 6–10.

The Race to Save the Lord God Bird by Philip Hoose (Farrar, 2004). Ages 12–15.

Thunder Birds: Nature's Flying Predators by Jim Arnosky (Sterling, 2011). Ages 8–11

What Bluebirds Do by Pamela F. Kirby (Boyds Mills, 2009). Ages 4–9.

INSECTS

At first glance, the number of citizen science projects devoted to bugs may seem surprising. Many are creepy-crawlies that aren't particularly well liked, after all, and others are small enough that we often don't pay them any notice. There are exceptions, of course. Butterflies are a beloved sign of spring, and they're well represented in citizen science projects. While it doesn't have the history of bird-watching, butterfly watching is a well-established hobby, thanks no doubt to the beauty and variety of butterfly species that can be seen. The idea of dragonfly watching as a hobby is a relatively new one. But it's up-and-coming, with a number of recently published field guides and newly formed clubs and websites dedicated to "odes"—short for *odonates*, the order that includes dragonflies and damselflies. And there's plenty left to learn, with a number of current citizen science projects aimed at just getting a handle on which dragonflies live where. The following projects offer opportunities to study these creatures and other insects in much more detail.

INSECTS
—
Ants

School of Ants

www.schoolofants.org

School of Ants is a study of ants that live in urban areas. Participants, which include significant numbers of school and family groups, collect ants in school yards and backyards to help create maps of which species live where. Participants make their own collection kits using index cards, cookies, and ziplock bags and use them to collect ants at four paved and four grassy locations on a single site. After registering their site, participants mail the specimens they collect to the School of Ants lab for identification and mapping. Collection requires about an hour of time.

The project is relatively new, but Andrea Lucky, head of the project, says that it is finding ant species that haven't been recorded in a location before. Sometimes that's a positive development, because it adds information about a species' native range. It can also be concerning, though. "One introduced species that we thought had been confined to the southwest has been found on the West Coast and in the upper Midwest," Lucky says. She and her team are now trying to determine the extent of the invasive species' expansion. They are also conducting genetic research to study population migrations, and investigating the effect of warming temperatures on ant species.

Collections can be made anywhere in the country. In 2012, the project also focused on Chicago, New York City, and Raleigh-Durham, North Carolina, seeking contributions from those regions in particular. "We get to ask more specific questions because we have more samples from a specific area at the same time," Lucky says. These regional pushes will continue in the future; check the project website for details on which cities will be included each year.

The project is suitable for all ages and can serve as a useful introduction to hands-on science. The idea that science is done in a lab with a white coat and goggles can blind us to the fact that we don't necessarily know much about the things we see every day. "There is science to observing natural history and tracking what we consider normal," Lucky says.

AGE RANGE All ages   **TIME FRAME** Year-round   **REGION** US   **FEE** None

BeeSpotter   http://beespotter.mste.illinois.edu

BeeSpotter is a citizen science project hosted at the University of Illinois to collect data about and photos of bees.

"The goals of BeeSpotter are to engage citizen scientists in data collection to establish a much-needed baseline for monitoring population declines, to increase public awareness of pollinator diversity, and enhance public appreciation of pollination as an ecosystem service," the project's website says.

Participants report their bee sightings along with photos, details about where and when the bee was seen, and the identity of any flowers the bees are spotted on. The project is currently limited to Illinois only, but organizers hope to expand to other states.

AGE RANGE All ages   **TIME FRAME** March through October   **REGION** Illinois
FEE None

The Great Sunflower Project   www.greatsunflower.org

Founded in 2008, the Great Sunflower Project is a citizen science project focused on bee conservation.

"It's important because studies have shown honeybee and native bee populations declining across the country," says Fred Bove, outreach director for the project. "We don't know how this is affecting the pollination of our gardens, crops, and wild lands. One-third of our food is touched by pollinators, so it's very important to find out."

Participants contribute by growing Lemon Queen sunflowers (or an alternate bee-attracting plant if the Lemon Queen is not available) and making regular fifteen-minute observations of those plants to count the bees that visit them throughout the blooming season. Participants can also observe wild plants. The project also encourages all of its one hundred thousand registered members to make a count on a specific day each year—August 11 in 2012—as part of the Great Bee Count.

Bove says that results have been variable around the country, although observations have shown that both large gardens and community gardens

have better-than-expected pollinator service, while urban areas generally report lower than average bee counts. All data is published on the project's website, and participants are also encouraged to contribute information to the USA National Phenology Network (see Nature's Notebook, page 163, for details).

Observing is suitable for all ages and can introduce students to the scientific process of making observations, recording them, and contributing them to a larger body of data. "As a society, we are becoming more concerned about children's ecological literacy," Bove says. "This gets them away from TV and video games and out observing nature."

AGE RANGE All ages   **TIME FRAME** Year-round   **REGION** US   **FEE** None

---

Native Buzz   www.ufnativebuzz.com

Native Buzz is a citizen science project created by the University of Florida Honey Bee Research and Extension Lab to study the nesting preferences and distribution of solitary bees and wasps—species of bees and wasps that live independently rather than in a hive, but that still are important pollinators.

"This project will help us by generating important data as to the nest preference and biodiversity of many other important pollinators [like bees] and biological control agents [like wasps]," says Jason Graham of the Honey Bee

Photo by Tyler Jones/UF/IFAS

Jason Graham examines a Native Buzz nest.

Research and Extension Laboratory at the University of Florida. He adds that bees and wasps face many problems, including colony collapse disorder, pathogens, and increasing pesticide use, and that it is also important to change public perception of bees and wasps from painful stinging insects to important contributors to the world's ecology.

Participants build nest sites for solitary bees and wasps out of wood, plastic, bamboo, paper, clay, or reeds. After registering their site with the project, participants collect information about when bees or wasps begin occupying a nest and when new bees or wasps emerge from it. Graham says that data will help to explain the nesting preferences of bees and wasps and learn their current status.

Participants can join as individuals or groups, and they can share photos or articles on the project's Facebook page. Graham adds that the project is easy for all ages to take part in and safe, as the species studied do not sting.

"Families can work together throughout the entire process," Graham says. "They can design and build together, and then set up a native buzz nest site right in their backyard or from a window in their home."

AGE RANGE All ages   **TIME FRAME** Year-round   **REGION** Worldwide   **FEE** None

---

WaspWatcher Program   www.cerceris.info

The WaspWatcher program actually seeks to protect trees by using observations of the crabronid wasp to monitor for the emerald ash borer, an invasive beetle that has the potential to devastate the population of ash trees in eastern North America.

The emerald ash borer is difficult to detect by visual surveys or traps. "Sniffing out alien beetles with native wasps has proven to be an effective approach to biosurveillance," the project's website says. The crabronid wasp preys upon beetles, including the emerald ash borer, and is native to the areas where the emerald ash borer has spread. By monitoring the wasps—which are relatively easy to identify and do not sting humans—project organizers hope to help find emerald ash borers so they can be controlled before they destroy trees.

Participants locate crabronid wasp nests in their area and monitor them to see what beetles the wasps feed upon. Specifically, they visit a nest on at least three

hot, sunny afternoons in July and August and collect fifty beetles from those nests to mail to the project's researchers.

AGE RANGE All ages **TIME FRAME** July and August **REGION** Ontario, New England, the Eastern Seaboard, Wisconsin, Ohio, and Michigan **$ FEE** None

---

BUTTERFLIES AND MOTHS

Butterflies and Moths of North America

www.butterfliesandmoths.org

The Butterflies and Moths of North America project is an independent effort to collect and provide access to quality-controlled data about butterflies and moths throughout the continent. Data comes from a variety of sources, including citizen scientists, professional lepidopterists, published literature, and museum and personal collections.

Citizen scientists can report butterflies and moths or their eggs, caterpillars, or cocoons or chrysalides through the BAMONA website. Contributors do not need to be able to identify the butterflies or moths they spot, but each submission must include a photo. About sixty expert volunteers review submissions and make identifications before adding the sightings to BAMONA's database.

About sixteen thousand people have registered for accounts to allow them to contribute their observations. They do so for a wide range of reasons: gardeners, retired people trying to document the animals on their property, and schools are regular users of the identification. In addition, "We get a lot of kids who find something cool and don't know what it is," says Kelly Lotts, BAMONA project coordinator. They or their parents will submit their photos for identification. It's particularly gratifying, she adds, when the submission comes from a parent who's creeped out by the bug. "A lot of times it can convert someone from being scared to someone who's excited to learn more," and that excitement can cascade to their children, Lotts says.

Scientists also use the information that the project has collected. Lotts says these uses tend to fall into three categories. First are researchers who are collecting all the information they can about a specific species. Second are those who are looking for widespread data to use for modeling and analysis. "Butterflies are an excellent indicator of climate change and shifts," Lotts says, since many species are dependent upon single plant species for their survival. Finally, some information requests come from land managers who want to know what species have been found locally so they don't take any actions that may threaten those animals.

INSECTS
—
Butterflies and Moths

AGE RANGE All ages   **TIME FRAME** Year-round
REGION North America   **FEE** None

Florida Butterfly Monitoring Network   www.flbutterflies.net

Established in 2003 by the McGuire Center for Lepidoptera and Biodiversity at the University of Florida, the Florida Butterfly Monitoring Network is a collaboration between scientists and volunteers to help protect Florida's butterfly populations.

Data generated by citizen scientists helps researchers and land managers facilitate strategic conservation planning, develop recovery or management actions, identify threatened butterfly species and critical remaining populations, track changes in butterfly populations, record vagrant or new species, identify potential environmental threats, and evaluate the impact of weather events such as drought or hurricane.

Volunteer responsibilities are based strongly on the requirements of the Illinois Butterfly Monitoring Network. They include six site monitoring visits per year, each of which takes one to two hours; continuing with the FBMN for multiple years; and learning to identify fifty to sixty butterfly species over the first two years of participation. New participants receive classroom and field training in butterfly biology and behavior, identification, and field monitoring and data collection methods. Review sessions are also available throughout the year.

AGE RANGE 18 and up   **TIME FRAME** Year-round   **REGION** Florida   **FEE** None

Illinois Butterfly Monitoring Network   www.bfly.org

Founded in 1987 by the Nature Conservancy, the Illinois Butterfly Monitoring Network utilizes volunteers to monitor the health of butterfly populations at more than one hundred sites throughout northeastern and central Illinois.

At the time of the project's founding, the state Nature Conservancy was overseeing a number of ecological restoration activities, primarily focusing on plants. IBMN Director Doug Taron says the network was founded to ask what effects those activities were having on animals. "Insects were chosen because they represent the majority of species diversity," he says, adding that butterflies were selected because they don't require specialized equipment to view, there are a reasonable number of species for amateurs to learn, and many are not well adapted to a changing landscape.

The network follows a scientific method called a Pollard Transect. Volunteers are assigned a fixed route to monitor and must commit to visiting their site at least six times between June 1 and August 7 each year. Site visits require one to two hours. Volunteers are expected to continue for multiple seasons and to learn to identify fifty butterfly species over their first two years.

Participants report the number of each type of butterfly they spot, broken down by the type of habitat they are found in. They also report the time of the observations and the weather conditions, since those variables can have a significant impact on what they see.

The IBMN is now housed at Chicago's Peggy Notebaert Nature Museum. It sends raw data that the project generates to the landowners of the monitoring sites, and a number of researchers use the data as well.

Taron says that most volunteers are over 50, female, and retired. But there are no age limits, and the project has had monitors as young as 9. Volunteering is perhaps less viable as a family activity than most citizen science projects. Site monitoring is a more or less solitary activity; a monitor can be accompanied by an assistant while touring his assigned site, but those assistants are permitted to help with record keeping and identification only—not spotting butterflies.

But while site monitoring isn't particularly social, the project does have a significant community aspect: the program provides training events for volunteers twice a year, and monitors form friendships with each other. "People enjoy the opportunity to get out and go into the field together," Taron says, even if those trips don't produce IBMN data.

The IBMN is a model for networks in several other states and regions; see Ohio Lepidopterists Long-Term Monitoring of Butterflies (page 97) and

Florida Butterfly Monitoring Network (page 91). Also, IBMN is currently seeking monitors for sites in northwest Indiana.

✝ AGE RANGE No limits, although it may not be well suited for young children **▦ TIME FRAME** June 1–August 7
⊕ REGION Northern Illinois and northeast Indiana **$ FEE** None

Michigan Butterfly Monitoring Network www.naturecenter.org

Founded in 2001, the Michigan Butterfly Monitoring Network is based out of the Kalamazoo Nature Center. "It had a history of birding, and that skill set fits well with butterflying," says Karen Wilson, the network's founder. Additionally, KNC had been doing some land conservation work, and needed an appropriate way to measure its impact on invertebrates.

"We literally took the Illinois Butterfly Monitoring Network protocol and applied it," Wilson says. That includes set routes visited several times per season to identify and count all butterflies seen.

Participants include a range of experience levels, although all are expected to learn to identify a large number of local butterflies early in their participation.

Wilson now works at the Peggy Notebaert Nature Museum in Chicago and with the Illinois Butterfly Monitoring Network. Michigan, Illinois, and other state networks across the country have been in contact with each other to share information and collaborate while retaining state-level autonomy. Those that follow similar protocols have been evaluating whether their methods are close enough that they could all contribute to a national database.

✝ AGE RANGE All ages, though best for adults **▦ TIME FRAME** June through August
⊕ REGION Michigan **$ FEE** None

Monarch Larva Monitoring Project www.mlmp.org

The University of Minnesota's Monarch Larva Monitoring Project collects long-term data on monarch caterpillar populations, as well as the milkweed that provides a habitat for them. "This data helps us to understand the

habitat available to monarchs and whether or not monarchs are using the habitat," says MLMP Community Program Assistant Wendy Caldwell.

Citizen scientists can contribute by making observations at sites where milkweed grows. Volunteers examine their sites to count monarch eggs and caterpillars and to identify which stage of larval development any caterpillars have reached. They also provide annual site descriptions and calculations of the site's milkweed density, and there are several additional optional activities. People who can't commit to regular monitoring or who don't have regular access to an appropriate habitat can also provide anecdotal observations of milkweed and monarch caterpillars, eggs, or butterflies when they see them.

"One of the key motivators for our volunteers is that they are involved in real scientific research," Caldwell says. Project observations have been used in research on a host of different topics, including the effect of herbicides in agricultural fields on milkweed, parasite rates in wild monarchs, and survival rates of monarch eggs and larvae.

It's also a project that is suitable for all ages. "Volunteers have told stories of monitoring with children as young as age 3," Caldwell says. Because there are several components to the citizen science observations, young children can start with the simplest observations and do more as they keep learning.

Participants can learn the necessary techniques from the project's website or by attending training sessions held throughout the country. By participating, Caldwell says, citizen scientists can learn about the monarch habitat, including the other plants and animals that use it, as well as the natural history of the site they monitor. "It is fun to find monarchs in the wild," Caldwell says. "Once they start monitoring for monarchs, most volunteers become hooked and spread the word to their friends, family, and neighbors."

AGE RANGE All ages   **TIME FRAME** Year-round   **REGION** North America
FEE None

Monarch Watch   www.monarchwatch.org

Founded in 1992, Monarch Watch is a program based at the University of Kansas dedicated to the conservation of the monarch butterfly and education

and research about it. "Monarch butterfly populations are declining due to loss of habitat," says Chip Taylor, Monarch Watch director, on the project's website. The animal's biology plays an important role: monarch caterpillars feed on the leaves of a single plant, milkweed. "To assure a future for monarchs, conservation and restoration of milkweeds needs to become a national priority," Taylor says.

Monarch Watch offers instruction and materials for citizen scientists to tag monarch butterflies, and maintains a database of tagged butterfly recoveries. Monarchs migrate up to three thousand miles each year to Mexico or the California coast. Tagging individuals helps scientists monitor these migration patterns.

The project is even responsible for the first monarchs in space: in 2009, three Monarch Watch caterpillars were aboard the space shuttle Atlantis. They completed their development and emerged as adult butterflies aboard the International Space Station.

Citizens can also help the project by creating and certifying Monarch Waystations—habitats with milkweed for the caterpillars to eat and nectar plants to provide food for adult butterflies. Seed kits are available for $16 from Monarch Watch.

AGE RANGE All ages   **TIME FRAME** August through October for tagging
REGION North America east of the Rockies **$FEE** Price for tagging kits varies from $15 for 25 tags to $105 for 500 tags

North American Butterfly Association Butterfly Counts
www.naba.org

The North American Butterfly Association is a membership-based organization working to conserve and increase public enjoyment of butterflies. As part of this, NABA collects data from formally organized butterfly counts throughout the year as well as free-form, anecdotal monitoring.

NABA built its butterfly count program using the Audubon Society's Christmas Bird Count as a model. It's organized into count circles, each of which covers a fifteen-mile radius. Each count circle is required to register with NABA, have at least four observers, and count for at least six hours.

There are nearly five hundred count circles throughout North America, and there's a lot of variation among them. While some circles have the minimum four people counting in a single preserve or state park, others can have up to nine or ten parties, each with four observers monitoring multiple areas.

Each circle makes its observations on a single day. Traditionally the counts have been organized in the weeks around Independence Days in North America: July 4 in the United States, July 1 in Canada, and September 16 in Mexico. Recently, NABA has added spring and fall counts in the United States. Individual circles can participate in all three of NABA's counts, although they are not required to and many do only one per year.

NABA data has been used in a number of scientific papers on climate change, land usage, and butterfly evolution and biology. The National Science Foundation, in fact, recently awarded the association a $1 million grant to help it improve the online distribution of its data so it can be more widely cited.

But the counts also serve a public awareness function. "We think the most important way to save butterflies and their habitats is to create constituents who care about the butterflies and their habitats," says NABA President Jeffrey Glassberg. Before the association was founded in 1992, Glassberg says, there were few active butterfly watchers in North America—and those who did watch butterflies also took nets to capture and kill them. Today, the ranks of butterfly watchers includes tens of thousands.

They are also social events. "You'll get to meet other people that like nature and butterflies," Glassberg promises. Family groups are particularly encouraged, and many counts have lots of children. Beginners are also welcome. "We often have people who are beginners and don't know how to identify butterflies," Glassberg says. "If they're a human being with two functioning eyes, they'll learn to recognize different species before long."

For individuals who can't join a count circle, NABA also hosts the "Butterflies I've Seen" database, where anyone can report butterfly sightings made under any circumstances. While not as formal as the count circles, Glassberg says the database still provides valuable data, which will become more useful over the next few years as the NSF grant is implemented.

AGE RANGE All ages; families particularly encouraged **TIME FRAME** Year-round; especially active in June and July (US and Canada) and September (Mexico) **REGION** North America **FEE** $3 per count circle participant defrays cost of compiling and publishing data

Ohio Lepidopterists Long-Term Monitoring of Butterflies

www.ohiolepidopterists.org/bflymonitoring

The Ohio Lepidopterists monitoring project is modeled on the Illinois Butterfly Monitoring Network. It provides information about fluctuations in population, changes in range, butterfly migrations, immigration of nonnative species, and overall butterfly health in the state.

Volunteers walk a fixed route at least once per week between April 1 and October 31. They record the number and type of butterfly they spot, broken down by the type of habitat each butterfly is spotted in. Volunteers also note any land management that takes place, such as mowing or cutting of stands of flowers, as well as weather data and information about any caterpillars spotted.

AGE RANGE All ages; families particularly encouraged **TIME FRAME** April 1 through October 31 **REGION** Ohio **FEE** None

Project Butterfly WINGS

www.flmnh.ufl.edu/wings

Project Butterfly WINGS (Winning Investigative Network for Great Science) is a citizen science project aimed at children ages 9–13 who track the presence and abundance of common butterfly species by state and county. The project has both research and educational goals. Participants perform long-term monitoring that helps collect information about butterfly ranges and phenology, but the project also encourages classroom-type inquiries into topics such as the food sources butterflies prefer or the effect of weather on butterfly activity.

Project Butterfly WINGS is a collaboration between the Florida Museum of Natural History, the Florida Cooperative Extension Service at the University of Florida, and National 4-H. Participants do not need to be part of a 4-H group, however.

AGE RANGE Primarily ages 9–13 **TIME FRAME** Year-round **REGION** US **FEE** A facilitator guide ($7.95) is required for participation

Monarch Alert Project

http://monarchalert.calpoly.edu

Operated by Cal Poly San Luis Obispo's biology depart-ment, the Monarch Alert Project samples populations of monarch butterflies overwintering in San Luis Obispo and Monterey Counties in California. Participants esti-mate the number of butterflies in roosting clusters at several sites.

Goals of the project include estimating the size of local overwintering populations; developing models to predict future population sizes; tagging individuals to study their movement and connections between overwintering sites; developing models to explain monarch tree use and movement between groves of trees; and creating scientific plans for land management, planning, and conservation.

AGE RANGE All ages   **TIME FRAME** October through February
REGION Monterey and San Luis Obispo Counties, California **FEE** None

Mothing

www.discoverlife.org/moth

Discover Life, an initiative of the Polistes Foundation, hosts the Mothing pro-gram to use citizen scientists to collect data and photographs of moths from a large number of sites. The program has twin goals of large-scale ecology research and science education.

Ecologically, "our goal is to look at how species are affected by large-scale factors, including climate, pollution, and invasive species," says Nancy Lowe, Discover Life outreach coordinator. Scientists don't know yet how climate change or extreme weather events affect moth species. "We can't assume it's all going to be good, and we can't assume it's all going to be bad," Lowe says. "We're scientists; we have to collect the data."

Participants photograph moths at a location of their own choosing and upload the pictures to the Discover Life website, along with documentation of the location. "We use photographs [rather than just observational notes] to allow expert identification," says Discover Life Founder John Pickering. "Therefore you can get very high quality data." Participants are welcome to identify species as well, but as there are twelve thousand moth species in North America, it can be a difficult task for amateurs. These identifications can be used to track life lists of species seen, and to create checklists of species expected to be seen at a specific site.

Moths are ecologically important animals: caterpillars provide food for other animals, while adult moths help pollinate plants. Mothing is also an easy project for all ages to get into, because it's convenient and the moths are attractive, can't bite, and provide an incredible variety of animals to observe. "It's really quite inspiring to see the diversity of creatures in just one group," Lowe says.

That makes it suitable as an introduction to natural history. "We're trying to train the next generation of naturalists," she says.

AGE RANGE All ages   **TIME FRAME** Year-round   **REGION** US   **FEE** None

ProjectMonarch Health   www.monarchparasites.org

Project MonarchHealth is a survey of OE, a single-cell parasite that infects monarch and queen butterflies. While it's difficult to find without careful examination, the parasite (full name: *Ophryocystis elektroscirrha)* can have serious impact. It's harmless to humans, but infected butterflies tend to be smaller and lighter than healthy butterflies, and can't fly as far or as long. In the worst cases, the butterfly will die before emerging from its chrysalis or before it can fully expand its wings. About 8 percent of eastern monarchs that migrate to Mexico each year have OE, but more than 70 percent of the non-migratory population in south Florida do.

The parasite is active while the butterfly is in its pupal (chrysalis) stage. When the butterfly is an adult, however, OE exists as microscopic spores on the butterfly's body. Participants in the projects sample wild monarchs, either by capturing adult butterflies or by collecting caterpillars and rearing them until adulthood. Using a monitoring kit supplied by the project, they test for OE by pressing a clear-tape sticker against the butterfly's abdomen, peeling it off, and attaching it to an index card. Participants also record data about each butterfly before returning the card to the Project MonarchHealth lab. Scientists at the program head Sonia Altizer's lab at the University of Georgia examine the cards under a microscope to determine if OE is present. Project participants sampled nearly ten thousand butterflies in twenty-seven states and two Canadian provinces between 2006 and 2010.

Butterflies and Moths

The project is open to all ages and skill levels. Its website includes a set of educational resources, including classroom activities, monarch-related PowerPoint presentations, and learning tools such as a glossary and crossword puzzles.

AGE RANGE All ages **TIME FRAME** April 15 through October 31 **REGION** North America **FEE** None

Southwest Monarch Study   www.swmonarchs.org

Based in Arizona, the Southwest Monarch Study is working on a question unique to the state. It's widely accepted that monarch butterflies east of the Rocky Mountains migrate south to Mexico each year, while monarchs west of the Rockies migrate to the coast of California. But Arizona is south of the Rockies. Where do *its* monarchs go?

Participants tag monarch butterflies in Arizona and monitor populations of the milkweed that serve as the food source for monarch caterpillars.

Project participants have tagged more than eight thousand monarchs in Arizona since 2007. Some of those tagged butterflies have been recovered in Mexico, some in California, and perhaps most interesting, some spend the winter near Phoenix or Yuma. "We always thought in the beginning that they would go to one place or another, not all three," says Gail Morris, coordinator of the study. She adds that the group plans to eventually publish a paper on the recovery data.

The study organizes trips to teach participants how to net and tag butterflies, and how to identify monarch caterpillars and the milkweed they eat. Participants can start tagging on their own once they've been on one of these trips.

Morris says the trips are well suited for families. "We're all inside so much I think we've lost that sense of wonder at the natural world," she says. "Families taking the time to do something like that with their kids is a gift that will come back many times."

Tagging butterflies is a good opportunity to learn about them and their role as an indicator of overall ecosystem health, Morris says. But the opportunity to be in nature has other benefits as well. "I don't know anyone that leaves tagging events stressed," she says. "They're seeing things in front of them in a new way. It keeps you humble because there's always something else to learn."

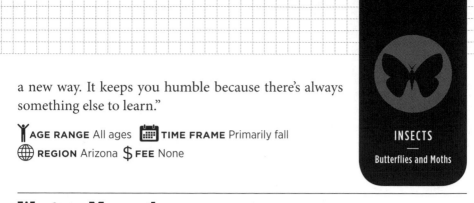

Y AGE RANGE All ages   📅 **TIME FRAME** Primarily fall
🌐 **REGION** Arizona $ **FEE** None

INSECTS
—
Butterflies and Moths

Western Monarch Thanksgiving Count

www.xerces.org/california-monarch-campaign

Monarchs that live west of the Rocky Mountains mostly migrate each winter to the California coast. In the 1980s, the Xerces Society, a nonprofit dedicated to invertebrate conservation, documented the specific places in California where monarchs overwinter. The Western Monarch Thanksgiving Count is a citizen science project that continues the monitoring of those sites to ensure that information is kept up to date.

"Having data from volunteers helps us to find what's out there for monarchs to use and the ideal monarch habitat," says Mía Monroe, coordinator of the California Monarch Campaign for the Xerces Society. The project also helps researchers determine the timing of major events in the monarch's life cycle, including breeding and migration.

Volunteers visit a number of monarch colonies each year, typically between 75 and 120 in total. Despite the name, participants are encouraged to survey monarch colonies several times per season, although the survey prioritizes observations made during Thanksgiving week and the week of January 1. In groups of at least two, participants visit colonies early in the day when temperatures are cool and the butterflies are less likely to fly. They estimate the number of butterflies roosting in specific trees and exhibiting other behaviors, and report weather data and details about the trees where butterflies are found and the height and form of the butterfly clusters.

"We can help communities take better care of their monarchs by noticing trends based on this information," Monroe says. All data is posted on the Xerces Society website, she adds, so the public can use it and so participants can see the impact that they are having. Dennis Fry, a retired professor of entomology, does scientific evaluation on the data "so when we post on the website we know there's science behind it," Monroe says.

Participating requires minimal skills or knowledge, but it offers the opportunity to see firsthand biological concepts such as the effect of weather on

behavior, the relationship between plants and animals, and the technical aspects of finding things in nature by eye or through binoculars. It's a popular activity for families or class groups, Monroe says: "Kids ask questions and see things adults don't."

Other people take part for the opportunity and the excuse to be in nature. As Monroe observes, "Monarchs are always in the most beautiful places."

AGE RANGE All ages **TIME FRAME** November through March **REGION** Coastal California from Mendocino to Mexico **$ FEE** None

---

DRAGONFLIES AND DAMSELFLIES

Dragonfly Pond Watch   www.xerces.org/dragonfly-migration/pondwatch

Dragonfly Pond Watch is the citizen science arm of the Migratory Dragonfly Partnership, which studies the fascinating but poorly understood migrations undertaken by about sixteen dragonfly species in North America. "Almost nobody knows that dragonflies migrate," says Celeste Mazzacano, project coordinator for the Migratory Dragonfly Partnership. "It's even a surprise to a lot of entomologists." This is despite the fact that dragonflies migrate on every continent except Antarctica and one, the Wandering Glider, can travel up to eleven thousand miles and cross the Indian Ocean.

Dragonfly Pond Watch focuses on five species. Participants visit a specific wetland or pond site regularly to observe and record when the first migrating dragonflies arrive in the fall and the spring, as well as when the first resident populations emerge at the site each spring.

Participants are asked to register their site and make observations at least once a month. They report the type and number of dragonflies at the site, as well as information about their sex, their stage of development, and their behaviors. Opportunistic (one-off) observations are also welcome.

There are plenty of questions that the project is investigating, including the environmental cues that trigger migration, where migrating dragonflies overwinter, how migrating dragonflies navigate, and the effect of climate change on migration patterns. In many cases, a single wetland will be home to both resident and migrant populations of the same species, and the project is also trying to determine if a migrant butterfly or its offspring ever become residents or vice versa.

Because participants need to learn to identify only five species, it's a relatively easy project for families to take part in. Each of the species is fairly common and widely distributed, so most people can find them in areas convenient to them. And it's a valuable contribution to science. "Dragonfly migration flights follow certain general pathways, but the timing and duration is so sporadic and unpredictable that these projects really need crowdsourcing—a wide array of people spread out across North America looking for migration," Mazzacano says.

INSECTS
—
Dragonflies
and Damselflies

▼ AGE RANGE All ages   **▦ TIME FRAME** Spring through fall   **⊕ REGION** North America
$ FEE None

The Dragonfly Swarm Project   http://thedragonflywoman.com/dsp

Entomologist Christine Goforth started the Dragonfly Swarm Project in 2009 while working at an aquatic ecology lab. Her duties included testing water quality at a lake once a week. One week, she arrived at the lake to discover hundreds of dragonflies swarming over the land near the lake. "It was terribly exciting because I hadn't seen it before," she recalls.

To satisfy her own curiosity, Goforth researched the phenomenon of dragonfly swarms but didn't find much information. "It's never really been studied because it's so hard to study," she says. "It happens often, but it's very localized, and conditions need to be perfect" for dragonflies to swarm.

Goforth also shared the story of the dragonfly swarm on her blog. While there had been little formal scientific inquiry into dragonfly swarms, plenty of readers were able to share their recollections of dragonfly swarms they had seen. From there, she began specifically asking for swarm reports and formally collecting the data. That data has been rolling in since: she recently received her two-thousandth swarm submission.

"Most of the people who report swarms think it's a major experience," Goforth says. "It touches their souls in a way they didn't expect."

Participants can report swarms through a form on the project's website. Information Goforth collects includes the date, time, and location of the swarm; whether it was a feeding or migratory swarm; weather conditions; and the number and type of dragonflies present.

"I'm trying to figure out the conditions that lead to swarms" so they are easier to predict and therefore study, Goforth says. Weather is a big factor, she has learned: storms can push the dragonflies around and together, while flooding creates a big concentration of the mosquitoes and other insects that dragonflies prey upon.

In fact, dragonflies' role as a predator of biting insects that are nuisances and that can spread disease means that studying them has potentially direct benefits to humans. "Understanding how these predators come into an area and take care of a population explosion is really valuable," Goforth says.

Goforth shares all of the information she receives on her blog, through weekly and annual reports. The project blog also includes links to dragonfly news and information. In the longer term, she plans to publish scientific papers based on the data after five years and to continue collecting information beyond that.

AGE RANGE All ages   **TIME FRAME** Year-round   **REGION** Worldwide   **FEE** None

---

Illinois Odontological Survey   www.illinoisodes.org

Modeled on the Illinois Butterfly Monitoring Network, IOS (formerly the Illinois Dragonfly Monitoring Network) performs much the same type of monitoring for odonates, the insect order that includes dragonflies and their smaller cousins, damselflies. Participation requires commitment to spend one to two hours walking a set route at least six times between late May and late September, as well as attending one spring workshop each year and learning to identify dragonfly and damselfly species.

Participants monitor relative densities of odonates by counting the number and type within a specific space and time. The technique detects long-term population trends, so participants are expected to continue with the project for multiple years.

AGE RANGE All ages   **TIME FRAME** May through September   **REGION** Illinois
FEE None

Manitoba Dragonfly Survey

www.naturenorth.com/dragonfly

INSECTS

Dragonflies and Damselflies

The Manitoba Dragonfly Survey was originally started by the provincial government to assess the diversity of dragonflies throughout Manitoba. It's now coordinated by *Nature North*, an online magazine about nature in the province.

Participants can contribute to the survey by sharing observations and photos of dragonflies throughout Manitoba, particularly in northern Manitoba, which has not been well surveyed. In some cases, participants are also invited to contribute voucher specimens to ensure proper identification and for future scientific reference.

The survey's goals include creating a detailed atlas and field guide, comparing dragonfly diversity in undisturbed habitats to human-managed ones, and to identify habitats of special conservation importance.

AGE RANGE All ages **TIME FRAME** Spring through fall; summer is peak
REGION Manitoba **FEE** None

Michigan Odonata Survey

http://insects.ummz.lsa.umich.edu/michodo/mos.html
http://mos-atlas.blogspot.com

Established in 1996, the Michigan Odonata Survey is a volunteer effort to document and catalog the dragonflies and damselflies that can be found in Michigan.

Participants can contribute by collecting odonate voucher specimens in target areas that are underrepresented in collections for deposition at the University of Michigan, collect data from museum collections, provide data from their own collections, and monitor for the appearance of targeted species.

Mark O'Brien and Julie Craves of the River Rouge Bird Observatory at the University of Michigan at Dearborn have begun work on an atlas of Michigan odonata based on the information gathered by the project. The atlas will serve as a detailed reference work rather than a field guide, with distribution maps, emergence records, and sections on the history and ecology of odonates in Michigan.

AGE RANGE All ages **TIME FRAME** Spring through fall **REGION** Michigan
FEE None

Minnesota Odonata Survey Project

www.mndragonfly.org

Minnesota serves as a crossroads for dragonflies and damselflies. Northern, southern, eastern, and western species all call the state home. "All directions converge in the state in pretty remarkable ways," says Kurt Mead, leader of the project.

Despite that, the Minnesota Odonata Survey Project's website notes that "Minnesota has the distinction of being one of the most unstudied states when it comes to odonates." There are counties that have no scientific records of dragonflies or damselflies, even though those insects certainly can be found in them. The Minnesota Odonata Survey Project seeks to rectify that and determine where in the state the various types of dragonflies and damselflies can be found, and how widely they are distributed.

"Before I started looking at dragonflies I had no concept of how many there were," Mead admits. When the project started in 2006, there were records of about 130 species in the state. By July 2012, that number had grown by more than 10 percent, to 144.

"We need a baseline to compare to as the climate shifts" and potentially changes the ranges of various species, Mead says.

Project participants report the types and number of dragonflies spotted, their gender, and their precise location. Participants may also photograph or collect specimens for the project to archive.

Unlike bird-watching, dragonfly watching is a relatively new hobby. Mead conducts four one-day workshops per year for new participants at locations across the state, as well as a weekend workshop each summer. These workshops are mostly in the field, where Mead and volunteers who have some experience are available to answer questions. "There's a fairly steep learning curve, but you can plateau fairly quickly," Mead says. The project also has an active Facebook page where people can share photos and stories and answer each other's questions.

It's also an activity that fairly young children can successfully participate in. "I encourage families to bring their kids, as long as they're not toddlers," Mead says. "Kids can develop a real appreciation for some of the overlooked critters." Youth is also a practical benefit to capturing dragonflies for examination, he says: "Hand nets to ten teens and ten adults, and I can tell you the teens are going to catch more bugs."

AGE RANGE All except toddlers **TIME FRAME** April through October; June and July are the peak
REGION Minnesota **$ FEE** None

Wisconsin Odonata Survey   www.wiatri.net/inventory/odonata/survey

Started in 2002, the Wisconsin Odonata Survey aims to document the ranges and important habitats of populations of dragonflies and damselflies in the state. Submissions are welcome from both professionals and amateurs and can take the form of observation reports, photographs, or collected specimens. Observations also require dates and specific locations.

"The goal of the survey is to increase knowledge of distinct and critical habitats of all of our 164 species throughout the state," says Bob DuBois, coordinator of the survey.

There is a learning curve involved in finding and identifying dragonflies, but it's rewarding for enthusiasts—and for children. "Kids are really interested in dragonflies. Learning about them is a great doorway to nature appreciation." DuBois recommends group outings, such as those offered by the newly formed Wisconsin Dragonfly Society, as an excellent way to get started.

Uncommon species are of particular interest to the survey. There are a number of odonate species that may have Wisconsin as part of their range but that have not been documented in the state or appear only in historical records. Many others have been sighted in the state, but scientists want more information about their habitats, ranges, or flight periods.

The project's website includes a range map for each species, and participants can see when a species is first documented in a county because that county will be colored on the range map. "Volunteers should get the feeling that they're contributing to something meaningful, because they are," DuBois says.

AGE RANGE All ages **TIME FRAME** April through October; June and July are the peak **REGION** Wisconsin **$ FEE** None

INSECTS
—
Ladybugs

The Lost Ladybug Project   www.lostladybug.org

The Lost Ladybug Project collects information about ladybugs, with particular attention paid to the rapid shift in populations of various ladybug species in different locations and why some species have experienced rapid population declines. "This is happening very quickly and we don't know how, or why, or what impact it will have on ladybug diversity or the role that ladybugs play in keeping plant-feeding insect populations low," according to the project's website.

Participants find and photograph ladybugs in their neighborhoods, and upload the images along with information about when and where they found the ladybugs.

More than three thousand contributors have reported more than seventeen thousand ladybugs throughout North America. Spottings include rare specimens, such as the nine-spotted ladybug, which is the New York state insect but thought to be extinct in the state until found in Suffolk County in 2011.

The project's website goes into detail about how to ensure the accuracy of data. It encourages participants to explore places they've already collected ladybugs from, in order to check that their observations were typical. It also explains how observers can help to improve the statistics, by reporting all of the ladybugs they find instead of just the rare species, or by reporting the number of people searching and the amount of time it took to find ladybugs to give a sense of the effort it took to find them, and by reporting every search that didn't find any ladybugs.

The site also includes significant information that makes it particularly suited for young children. These resources include books divided by audience age, lesson plans, and a dedicated kid's page.

AGE RANGE All ages; the project is particularly well suited for elementary-age children
TIME FRAME Primarily early summer **REGION** North America **$ FEE** None

Resources

ADULT BOOKS

Attracting Native Pollinators: Protecting North America's Bees and Butterflies by Eric Mader, Matthew Shepherd, Mace Vaughan, and Scott Black (Storey Publishing, 2011).

Bees of the World by Christopher O'Toole and Anthony Raw (Facts on File, 2004).

The Bees of the World by Charles D. Michener (Johns Hopkins University Press, 2007).

Bees, Wasps, and Ants: The Indespensable Role of Hymenoptera in Gardens by Eric Grissell (Timber Press, 2010).

Butterflies of North America by Jim Brock and Kenn Kaufman (Houghton Mifflin Harcourt, 2006).

Butterflies of Ohio by Jaret Daniels (Adventure Publications, 2004).

Butterflies through Binoculars: The East by Jeffrey Glassberg (Oxford University Press, 2001).

Butterflies through Binoculars: Florida by Jeffrey Glassberg (Oxford University Press, 2000).

Butterflies through Binoculars: The West by Jeffrey Glassberg (Oxford University Press, 1999).

Caterpillars in the Field and Garden by Thomas Allen, Jim Brock, and Jeffrey Glassberg (Oxford University Press, 2005).

Caterpillars of Eastern North America by David Wagner (Princeton University Press, 2005).

Chasing Monarchs: Migrating with the Butterflies of Passage by Robert Michael Pyle (Mariner, 2001).

Common Dragonflies of the Southwest: A Beginner's Pocket Guide by Kathy Biggs (Azalea Creek, 2004).

Damselflies of the North Woods by Bob DuBois (Kollath-Stensaas, 2005).

Damselflies of the Northeast by Ed Lam (Biodiversity Books, 2004).

A Dazzle of Dragonflies by Forrest Mitchell and James Lasswell (Texas A&M University Press, 2005).

Dragonflies and Damselflies of Georgia and the Southeast by Giff Beaton (University of Georgia Press, 2007).

Dragonflies and Damselflies of Texas and the South-Central United States: Texas, Louisiana, Arkansas, Oklahoma, and New Mexico by John C. Abbott (Princeton University Press, 2005).

Dragonflies and Damselflies of the East by Dennis R. Paulson (Princeton University Press, 2011).

Dragonflies and Damselflies of the West by Dennis R. Paulson (Princeton University Press, 2009).

Dragonflies of the North Woods by Kurt Mead (Kollath Stensaas, 2009).

Dragonflies through Binoculars by Sidney W. Dunkle (Oxford University Press, 2000).

The Family Butterfly Book by Rck Mikula (Storey Publishing, 2000).

The Forgotten Pollinators by Stephen Buchmann and Gary Paul Nabhan (Island Press, 1997).

How Not to Be Eaten: The Insects Fight Back by Gilbert Waldbauer (University of California Press, 2012).

Keeping the Bees: Why All Bees Are at Risk and What We Can Do to Save Them by Laurence Packer (HarperCollins, 2010).

Mariposa Road: The First Butterfly Big Year by Robert Michael Pyle (Houghton Mifflin Harcourt, 2010).

Mason Bees for the Backyard Gardener by Sherian Wright (Inkwater Press, 2010).

Milkweed, Monarchs, and More: A Field Guide to the Invertebrate Community in the Milkweed Patch by Ba Rea, Karen Oberhauser, and Michael Quinn (Bas Relief, 2011).

Pollinator Conservation Handbook by Matthew Shepherd, Stephen Buchmann, Mace Vaughan, and Scott Black (Xerces Society, 2003).

Sex on Six Legs: Lessons on Life, Love, and Language from the Insect World by Marlene Zuk (Houghton Mifflin Harcourt, 2011).

Stokes Beginner's Guide to Dragonflies by Blair Nikula, Jackie Sones, Donald Stokes, and Lillian Stokes (Little, Brown, 2002).

CHILDREN'S BOOKS

Ant Cities by Arthur Dorros (HarperCollins, 1988). Ages 4–7.

Are You a Dragonfly? by Judy Allen (Kingfisher, 2004). Ages 5–7.

Are You a Ladybug? by Judy Allen (Kingfisher, 2000). Ages 5–7.

Butterflies by Seymour Simon (HarperCollins, 2011). Ages 6–10.

Butterflies and Moths by Nic Bishop (Scholastic, 2009). Ages 7–9.

The Buzz on the Bees: Why Are They Disappearing? by Shelley Rotner (Holiday House, 2010). Ages 5–7.

OTHER
ANIMALS
—

Caterpillar to Butterfly by Laura Marsh (National Geographic, 2012). Ages 8–12.

Creep and Flutter: The Secret World of Insects and Spiders by Jim Arnosky (Sterling, 2012). Ages 7–10.

Dazzling Dragonflies: A Life Cycle Story by Linda Glaser (Millbrook, 2008). Ages 6–9.

Dragonflies by Margaret Hall (Capstone, 2008). Ages 5–7.

Dragonflies of North America: A Color and Learn Book with Activities by Kathy Biggs (Azalea Creek, 2007). Ages 4–6.

Dragonfly's Tale by Kristina Rodanas (Clarion, 1992). Ages 7–9.

The Honey Makers by Gail Gibbons (HarperCollins, 1997). Ages 6–8.

How to Raise Monarch Butterflies: A Step-by-step Guide for Kids by Carol Pasternak (Firefly, 2012). Ages 7–10.

The Insect Detective by Charlotte Voake (Candlewick, 2010). Ages 3–6.

Monarch and Milkweed by Helen Frost (Atheneum, 2008). Ages 4–6.

Monarch Butterfly by Gail Gibbons (Holiday, 1989). Ages 6–8.

Not a Buzz to Be Found: Insects in Winter by Linda Glaser (Millbrook, 2011). Ages 7–10.

Sunflower House by Eve Bunting (Houghton Mifflin Harcourt, 1999). Ages 4–6.

OTHER ANIMALS

Not a fan of bugs, birds, or snakes? Plenty of other creatures have specific, dedicated citizen science projects.

Front Range Pika Project
www.pikapartners.org

The Front Range Pika Project is a citizen science program created by Rocky Mountain Wild and the Denver Zoo that collects data about the American pika, an animal related to rabbits that lives in the mountains of western North America, often above the tree line.

"Our long-term monitoring program is designed to gain an understanding of pika distribution and improve the long-term viability of this vulnerable species," the project's website says. Because the American pika's habitat is so remote, little is known about it. It's also highly sensitive to heat, and temperatures above

80 degrees for more than a few hours can be lethal. As a result, the American pika might be an early indicator of biological response to climate change.

"Pika Patrol volunteers follow monitoring protocols to collect data about pikas and their habitat in high altitude field sites," the project's website says. Observations include information about habitat, weather conditions, presence of water, and size of nearby rocks.

AGE RANGE All ages **TIME FRAME** Year-round **REGION** Rocky Mountains **FEE** None

---

JellyWatch   www.jellywatch.org

JellyWatch is a site for citizen scientists to report jellyfish, red tide, squid, or other unusual marine life. While an important part of the ocean's food web, increases in jellyfish populations may have adverse impacts on that food web, as well as on industry and tourism that relies upon the ocean.

"I started the project because there's a real lack of information" about jellyfish blooms, says Steve Haddock, scientist at the Monterey Bay Aquarium Research Institute. Conventional wisdom suggests that that the population of jellyfish and the frequency of jellyfish blooms where large groups gather in a small area are increasing, and that those situations reflect that the ocean is out of balance due to climate change or overfishing. "But there are very few time series that would measure that," Haddock says, so in reality we don't know if jellyfish populations have grown.

"In order to figure out what's happening with the ocean, we wanted to use a distributed citizen science approach," Haddock says. "This is kind of an ideal problem for the citizen science band of observers out there," since researchers can't possibly be everywhere the jellyfish might be to collect data.

Anyone can contribute observations of jellyfish on beaches or in the oceans whenever they see them, ideally with photographs to document and allow for expert identification. Participants can also report clean seas when they spend time on the beach or the water and find no jellyfish. The program is opportunistic, although Haddock says that several participants do make regular reports from their local beaches. Scientists can piece together reports from around a region to identify unusual jellyfish activity.

Data collected are freely available to anyone interested, and Haddock says that several researchers have used it in their investigations of specific jellyfish species. Open-water swimmers also use the database to determine if their routes are likely to be safe for swimming.

In addition to getting a better sense of whether jellyfish populations are changing, Haddock hopes the project will help to "educate people and getting them to understand the different kinds of jellies and why blooms happen." He also wants to provide some balance to the sensationalized way that jellyfish can be portrayed in the media. While alarmist news accounts may suggest that jellyfish blooms are unprecedented, they are natural events with fossil records dating to prehistory.

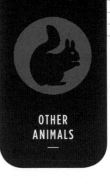

OTHER ANIMALS

AGE RANGE All ages **TIME FRAME** Year-round **REGION** Worldwide **FEE** None

New York Horseshoe Crab Monitoring Network

www.nyhorseshoecrab.org

Participants in the New York Horseshoe Crab Monitoring Network take observations about horseshoe crabs in New York's Marine District on or near Long Island. "Participants assist with the collection of scientific data that is used to assess the status of horseshoe crabs in New York State, and will help determine the management and conservation of this important species throughout the region," the project's website says. The project was developed by Cornell University Cooperative Extension's Marine Program and the New York State Department of Environmental Conservation.

Monitoring protocols vary by site, but generally include counting horseshoe crabs in a section of beach, counting the numbers of each sex, and tagging crabs so they can be identified if seen again. Monitoring takes place on evenings with full moons or new moons in May, June, and July when the crabs are spawning.

Prospective participants can select the beach they want to monitor from several survey sites. Site coordinators provide in-person training.

A free iPhone app is available as a tool for citizen scientists or casual beachgoers to record their horseshoe crab observations.

AGE RANGE All ages; children under 18 must be accompanied by parent or guardian
TIME FRAME May through July **REGION** Long Island, New York **FEE** None

Project Squirrel

www.projectsquirrel.org

Project Squirrel is a nationwide census of gray squirrels and fox squirrels sponsored by the Peggy Notebaert Nature Museum and the University of Illinois at Chicago. Participants can count squirrels around their home, office, school, or any other venue and report them through the project's website.

"One reason we chose squirrels is because squirrels are the most bona fide connection most people have with a wild animal," says Project Squirrel Coordinator Steve Sullivan. "Tree squirrels have a place in our collective psyche."

But as common as squirrels are, there are plenty of interesting questions about them to investigate. "Empirically, we know that the two species could live together, but they generally don't," Sullivan says. In Chicago, the two species sometimes even obey political divisions, which raises the question of whether human factors such as differences in garbage collection, dog laws, or rat control measures might be the cause.

Project Squirrel participants report information about the number and type of squirrels they see, as well as the location, the nearby tree species, the squirrels' food source, and what other animals they observe at the site. Participants can observe as often as they like, although the project does hope they will count squirrels at the same site at least once per season. Observing requires only a few minutes.

"You don't need to be a specialist," Sullivan says. "You just need to get out and have fun."

Those who want a more in-depth experience can also create food patches for squirrels and conduct foraging experiments in their yard. Even this requires minimal infrastructure, however, and the experiments are easy for kids to conduct with adult supervision. "For a family that wants to be involved

Become a Citizen Scientist.
Tell us about the squirrels near you at:

ProjectSquirrel.org

Some towns have one species of squirrel, others have two. By reporting your observations throughout the year, you can help us learn about the health of our environment.

Gray Squirrel *Sciurus carolinensis*

Fox Squirrel *Sciurus niger*

The promotional flyer for Project Squirrel doubles as a bookmark.

in scientific experimentation, I think it's kind of appealing," Sullivan says.

⫟ AGE RANGE All ages **▦ TIME FRAME** Year-round
⊕ REGION US and Canada **$ FEE** None

Whale Song Project

www.whale.fm

Whales and dolphins communicate with one another through sophisticated calls, and closely related individuals share calls and dialects. The Whale Song project from Zooniverse seeks to use citizen scientists to help analyze the calls of killer whales and pilot whales that have been recorded in order to understand how these animals communicate with one another and to understand the effect of human-generated sounds on them.

"The communication of killer whales and pilot whales is still poorly understood," the project's website says. "While we know for some species the general context in which sounds are made, many of the calls remain a mystery to us."

Participants work from their computers rather than in the field. Participants compare given segments of a whale's call to a set of potential matching calls to identify the most similar ones. Researchers hope that will help answer questions about the size of whale call repertoires, whether long- and short-finned pilot whales have different call dialects, and how well volunteers can categorize whale calls.

⫟ AGE RANGE All ages **▦ TIME FRAME** Year-round **⊕ REGION** Worldwide **$ FEE** None

WormWatch

www.wormwatch.ca

WormWatch is a citizen science monitoring program that collects data about earthworm species and habitats.

"The number of worms in a specific volume of earth can tell us a lot about how the habitat is being managed, because earthworms are very sensitive to soil disturbance," the project's website says.

Participants select their own sites and can choose from three protocols to count either worms living underneath objects or within the soil. Participants count worms, cocoons, juveniles, and worms that are estivating (resting in a

OTHER ANIMALS
—

coiled knot during summer to wait for more favorable conditions). They also identify worm species.

WormWatch is often used in classrooms. It offers teachers the opportunity to give real-world examples in their science classes, says Marlene Doyle, NatureWatch manager. The project's website offers a field guide to earthworms as well as an overview of earthworm ecology and anatomy.

Ÿ AGE RANGE All ages **▦ TIME FRAME** Year-round **⊕ REGION** Canada **$ FEE** None

Resources

ADULT BOOKS

Among Giants: A Life with Whales by Charles Nicklin (University of Chicago, 2011).

Horseshoe Crab: Biography of a Survivor by Anthony Fredericks (Ruka Press, 2012).

Horseshoe Crabs and Velvet Worms: The Story of the Animals and Plants that Time Has Left Behind by Richard Fortey (Knopf, 2012).

Mammals of North America by Roland Kays and Don Wilson (Princeton University Press, 2009).

North American Tree Squirrels by Michael Steele and John Koprowski (Smithsonian, 2003).

Peterson Field Guide series (Houghton Mifflin, various).

A Sand County Almanac by Aldo Leopold (Oxford University Press, 2001).

Squirrels: The Animal Answer Guide by Richard Thorington Jr. and Katie Ferrell (Johns Hopkins University Press, 2006).

The Whale: In Search of the Giants of the Sea by Philip Hoare (Ecco, 2010).

CHILDREN'S BOOKS

About Crustaceans by Cathryn Sill (Peachtree, 2004). Ages 4–6.

Jellyfish by Louise Spilsbury (Heinemann, 2010). Ages 5–8.

Killer Whales by Kate Riggs (Creative Education, 2012). Ages 6–9.

Seashells, Crabs and Sea Stars by Christinan Tibbitts (Paw Prints, 2008). Ages 7–10.

Squirrel Kits by Ruth Owen (Bearport, 2011). Ages 4–6.

Squirrels by Julie Lundgren (Rourke, 2010). Ages 5–8.

Whales by Tom Greve (Rourke, 2011). Ages 6–9.

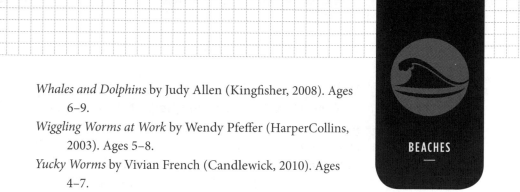

Whales and Dolphins by Judy Allen (Kingfisher, 2008). Ages 6–9.

Wiggling Worms at Work by Wendy Pfeffer (HarperCollins, 2003). Ages 5–8.

Yucky Worms by Vivian French (Candlewick, 2010). Ages 4–7.

Beaches, Wetlands, and Waterways

Ocean beaches are recreational hotspots, but they're also highly ecologically valuable sites, providing animal habitat and valuable indicators about the health of the ocean that covers 70 percent of the planet. Several citizen science projects have been launched to monitor beaches on both coasts to help ensure the health of this vital ecosystem. Rivers and wetlands are critical animal habitats too, and are often vital for human health. The following projects represent efforts to ensure the stability of wetland habitats.

BEACHES

Beach Watch   www.farallones.org/volunteer/beach_watch.php

Beach Watch is a long-term shoreline monitoring project in the Gulf of the Farallones and Monterey Bay National Marine Sanctuaries. It has collected information about the conditions of the beaches, including live and beached animals and the presence of tar balls or beach wrack, for twenty years.

"When we have events, whether natural die-offs or human-related events, it's hard to tell the effect on natural resources without baseline data," says Kirsten Lindquist, manager of Beach Watch. The information that Beach Watch has collected has been used to help garner major settlements in a number of oil spill cases, including the fifty-thousand-gallon Cosco Busan spill in 2007. "We've trained volunteers in the proper method of taking data so the evidence can be taken to court," says Mary Jane Schramm, media and public outreach specialist for the Gulf of the Farallones National Marine Sanctuary.

BEACHES
—

Data collected by the project is also used to evaluate waste management practices, monitor animal species of special concern, and identify potential areas for wildlife protection designation.

Volunteers survey an assigned beach once per month, often in teams of two or three. Each beach has two teams, so data about its conditions is collected every other week. "Our Beach Watch volunteers go out in very challenging conditions," says Schramm. "They hit the beaches religiously. They get a sense of ownership that a casual beach visitor won't get."

Volunteers must be at least 18, pass an application process, and commit to monthly surveys for at least one year. They receive about eighty hours of classroom and field training. "People come away with great identification skills," Lindquist says.

Data is also examined by a number of researchers in a variety of fields. "Volunteers take photos of the beach profile, so we have ongoing profile of how the beach changes," says Schramm, which will give an important indication of the effect of climate change.

AGE RANGE Adults **TIME FRAME** Year-round **REGION** Monterey Bay and the Gulf of the Farallones, California **$ FEE** None

---

British Columbia Beached Bird Survey

www.bsc-eoc.org/volunteer/bcbeachbird

Bird Studies Canada operates the British Columbia Beached Bird Survey to collect information on the causes and rates of seabird mortality on the British Columbia coast. "British Columbia's coastal waters support some of the highest densities of seabirds, waterfowl, and shorebirds in the eastern North Pacific," notes the project's volunteer manual. But because the coast also has significant shipping traffic, pollution from oil spills, fuel, and bilgewater are threats to these birds.

Volunteer surveys collected baseline data from 1989 to 1993. After a hiatus, the program was restarted in 2002.

Participants survey their beaches during the last week of each month year-round. In addition to the birds they find, volunteers report weather conditions, physical conditions of the beach, and the type and amount of oil pollution they find.

AGE RANGE All ages   **TIME FRAME** Year-round
REGION Coastal British Columbia   **FEE** None

Cape Cod Citizen Science Phenology Monitoring

www.nps.gov/caco/naturescience/phenology-monitoring-program-resources.htm

In 2011, Cape Cod National Seashore implemented a citizen science phenology monitoring program that provides some phenology observations complementary to the Cape Cod Long Term Ecosystem Monitoring Program, which is nested within the National Park Service Inventory and Monitoring Program. Phenology is the study of plant and animal life cycle events and how their timing is affected by climate variations. Project volunteers make observations of several specific plants and animals.

Participants make phenology observations one to two times per week between December and July; individual sites have shorter monitoring periods depending on the organisms found there.

The project intends to monitor phenology over the long term, in hopes that understanding local phenology will improve local management decisions. "We are interested in the status of the natural resources we manage," says Megan Tyrrell, research and monitoring coordinator at Cape Cod National Seashore. "When interest surged about climate change and involving people in science it seemed like a good fit to have citizen scientists collect information."

Most of the information volunteers collect is also shared with the USA National Phenology Network. With the exception of ice cover and salt marsh plant biomass observations, which are specific to the park, the Cape Cod program was designed to fall under the USA-NPN umbrella, says Scott Buchanan, science communication technician.

While there are no restrictions on who can participate, many participants are retirees who are in Cape Cod year-round. "I would guess most of are volunteers are part of the natural history guild or retired naturalists or teachers," Tyrrell says.

Buchanan adds that they do need to be dedicated to monitor regularly, as well as interested in science, natural resources, and conservation.

AGE RANGE All ages **TIME FRAME** December through July
REGION Cape Cod National Seashore, Massachusetts
FEE None

Coastal Observation and Seabird Survey Team

http://depts.washington.edu/coasst

A project of the University of Washington in partnership with state, tribal, and federal agencies, COASST is a seabird monitoring program at nearly six hundred beaches from Northern California to the Arctic Circle in Alaska.

The program was founded in 1998 to collect baseline data that would help to evaluate the effects of natural and man-made events—particularly oil spills—on seabird mortality. Participants survey a beach monthly for dead birds that wash up on shore. Volunteering requires a single six-hour training session where volunteers learn the survey protocols and to identify beached birds.

"They're out collecting data in a standardized fashion," says COASST Executive Director Julia Parrish. "That means those data are immensely useful to science."

The information collected can even be useful in ways no one anticipated. Parrish says one anthropology student was studying the people of southwest Washington from thousands of years ago, and found the bones of shearwaters—seabirds that normally wouldn't be found close to shore—in their garbage. COASST data showed that shearwaters occasionally wash ashore in great numbers, and the carcasses are usually fresh, suggesting that the birds were likely a food source for coastal peoples. "I never would have designed a citizen science program to answer that question, but it's really nice that the data can," Parrish says.

The program is well suited for families, which make up about 20 percent of COASST's more than eight hundred volunteers. It's also potentially a social activity. Surveys are usually conducted in pairs, and there is a lot of communication among participants.

Enough, in fact, for romance: there have been three marriages among COASST volunteers.

⛹ AGE RANGE All ages   📅 TIME FRAME Year-round
🌐 REGION Pacific beaches from northern California to Alaska
💲 FEE None, but a small deposit is required for survey and training supplies

BEACHES
—

Coastal Ocean Mammal and Bird Education and Research Surveys

www.sanctuarysimon.org/monterey/sections/beachCombers

Known as Beach COMBERS, this collaboration among Moss Landing Marine Laboratories, the California Department of Fish and Game Office of Spill Prevention and Response, Monterey Bay National Marine Sanctuary, the University of California at Davis Wildlife Health Center, and other state and research institutions in California uses citizen scientists to collect information about beached marine birds, mammals, and sea turtles at beaches in central California.

"It's nice going out on the beach, but it's really great to be able to contribute to greater understanding of coastal and marine ecosystems," says Hannah Nevins, project leader.

Project goals include collecting baseline information on beached marine animals, determining the causes of seabird and marine mammal deaths, helping detect natural and man-made events that kill birds and mammals, and assessing the abundance of tar balls and marine debris on beaches. The information generated has been used in a number of scientific papers and research reports, as well as shaping public policy. The project helped to prove that a gillnet fishery was killing large numbers of seabirds, which forced it to change its practices. It's also useful for working with the media. "When there's something unusual on the beaches, we get called," Nevins says. "Because we have rigorous data, we can inform about what's going on in a really timely way."

Monitoring is often done on relatively remote beaches. "It's an opportunity to get to a site where not many people go," Nevins says. She adds that volunteers often become passionate advocates for their beaches, which helps to inspire beach cleanups and Styrofoam bans.

BEACHES
—

Participants receive at least twenty hours of training and commit to a minimum of twelve surveys. They are expected to monitor a segment of beach up to five kilometers long during the first week of each month, requiring about four hours of work. Participants generally monitor in teams of two, although companions who do not collect data are welcome to accompany them. As a result, "it's a good way to meet others with similar interests," Nevins says, adding that partners frequently build a strong sense of camaraderie.

AGE RANGE Primarily adults; some high school children accompany parents
TIME FRAME Year-round **REGION** Monterey Bay, Cambria/San Simeon, Morro Bay, Santa Barbara, and Ventura, California **$ FEE** None

Long-Term Monitoring Program and Experiential Training for Students

www.limpetsmonitoring.org

LiMPETS is a monitoring and education program for students and volunteer groups on California's coast coordinated by the NOAA Office of National Marine Sanctuaries. Participants provide baseline monitoring of intertidal habitats—the section of coast below water at high tide but above it at low tide—to improve understanding of them and to help assess the impact of events such as oil spills.

Group leaders must be trained by a local coordinator. "We want to ensure they've been trained in invertebrate and algal species so they're making accurate observations," says Claire Fackler, NOAA national education liaison. After the leader has been trained, the group chooses an established site to monitor. On rocky beaches, participants make a number of observations, including identifying the species found at specific points, identifying the species found at a number of random places within a specific area, counting all individuals of a certain species within an area, and measuring the size of owl limpets in a certain area.

On sandy beaches, the project is primarily concerned with monitoring the density and distribution of the Pacific mole crab over time. Participants collect sand core samples at specified locations and examine them to see if mole crabs are present and to determine their size and gender.

"Participants get an authentic, hands-on coastal monitoring experience," Fackler says. The opportunity to get their hands and feet wet doing science can help to build excitement and give them something to share with friends and family.

Groups can monitor as often as they wish. Multiple groups monitor at each site to ensure regular observations. Because training is required, interested groups should first contact a regional coordinator through the project's website.

AGE RANGE Middle school and up **TIME FRAME** Year-round **REGION** Coastal California **FEE** None, although groups need to make, purchase, or borrow equipment, which can cost up to $250

Puget Sound Seabird Survey   www.seattleaudubon.org

Organized by Seattle Audubon, the PSSS is a survey of seabirds at specific locations on the first Saturday of the month from October through April.

"The survey is synchronized to take place during the same four-hour window, resulting in a simultaneous snapshot of seabird density on more than 2,400 acres of nearshore saltwater habitat," says Marieke Stientjes Rack, Seattle Audubon volunteer coordinator. Participants count birds on the water within three hundred meters of their location, as well as noting the species, weather conditions, activity of humans and raptors in the area, and oiling rates. That information will be used to develop long-term baseline data about seabird density that will function as an indicator of the environmental health of Puget Sound.

Data can be viewed by anyone on the project's website. It is also analyzed by the PSSS Working Group to examine trends and quantity the effects of environmental events such as oil spills. The Washington Department of Fish and Wildlife has incorporated Survey data into its oil spill response planning.

"Analysis of the first three years of data indicate that the majority of seabird densities continue to decline within Puget Sound," Rack says. Some declines are the result of natural factors, including El Niño and La Niña events that affect food availability, but others reflect human influences.

BEACHES
—

Volunteers receive training from Seattle Audubon before their first session. In general, surveys are conducted in pairs or teams.

Rack says that the project is not ideal for families with young children. "This survey uses advanced to expert birders, and relies on the full and complete attention of participants," she says. Seattle Audubon's Neighborhood Bird Project accommodates birders of all levels and is more flexible for families.

Save Our Shorebirds   www.mendocinocoastaudubon.org/mcas_cons.html

Save Our Shorebirds started five years ago when birders and biologists in the Mendocino Coast of Northern California noticed a decline in coastal birds. The citizen science project seeks the reason for the decline, and ultimately aims to reverse it.

Organized by Mendocino Coast Audubon Society and California State Parks, volunteers monitor watch-listed shorebirds on three beaches in MacKerricher State Park. They also remove invasive weeds from shorebird habitats.

AGE RANGE All ages   **TIME FRAME** Year-round   **REGION** Northern California **FEE** None

Seabird Ecological Assessment Network   www.tufts.edu/vet/seanet

Often shortened to SEANET, the Seabird Ecological Assessment Network conducts beached bird surveys along the East Coast of the United States to provide baseline information about bird deaths and to help detect natural and human-caused events that kill marine birds. The project was created in 2002 by the Tufts Center for Conservation Medicine.

"Marine birds can serve as indicators of ecosystem and human health," the project's website says. "Monitoring the threats they face and their mortality patterns can teach us about the health of the marine environment."

Volunteers commit to walking a section of Atlantic coastline at least one kilometer long every month or more often. They collect information on live and beached birds they see, as well as weather, beach conditions, and the presence of oil or litter.

AGE RANGE All ages   **TIME FRAME** Year-round
REGION US East Coast   **$ FEE** None

Citizen Monitoring Research: Freshwater Sponges

http://dnr.wi.gov

Freshwater sponges can be found in some of Wisconsin's lakes and rivers. To help study them, Wisconsin's Department of Natural Resources created a citizen monitoring program in 2007 through which members of the public can report sponges they see.

"It's an opportunity for people who spend a lot of time outdoors—or who just are out on a lake once a year—to look for something they might not know is out there," says project coordinator Dreux Watermolen.

Freshwater sponges are important parts of aquatic ecosystems, and their absence can be an indicator of pollutants. "Freshwater sponges have been studied more in Wisconsin than maybe any place in the world," Watermolen says, although they are still not well understood. Existing research includes baseline data collected in the 1930s and '40s and a study that suggested declines of some species in the 1990s, but "we really don't have a good handle on where freshwater sponge species occur," Watermolen says.

The project is opportunistic, so participants can report observations whenever they see them. "A lot of folks have questions about what they see," Watermolen says, and when they report their sightings, the project can help answer them.

No observation method is prescribed, although participants searching for sponges generally need to wade in shallow water and examine rocks, logs, and sticks where sponges might grow. Participants report the location and date of their observation, the material the sponges were growing on, and whether the sponges all appeared to be the same type. Citizen science observations

will be used to prioritize lakes for future research, which Watermolen hopes will have a citizen science component as well.

AGE RANGE All ages **TIME FRAME** Late summer and early fall **REGION** Wisconsin **FEE** None

Gastineau Guiding Company www.stepintoalaska.com

Gastineau Guiding Company offers scenic tours near Juneau, Alaska, aimed primarily at the nearly one million visitors who come to the city by cruise ship each year.

In a 2009 meeting with the Holland America cruise line, Gastineau Guiding developed the idea for a shore excursion that would give something back to the environment. The company's Alaska's Whales and Science Adventure offers the same sights and wildlife as the rest of the company's offerings, but it also incorporates a citizen science component. "It offers people a chance to see the science behind the scenery," says Jeremy Gieser, the company's director of tours and marketing.

While the half-day tours are, by necessity, limited to scientific basics, their impact is potentially lifesaving. Part of the tour takes place on the water in Stephens Passage, where citizen scientists monitor harmful algae blooms that can cause paralytic shellfish poisoning. Shellfish that contract the toxin can be lethal for human consumption, but Gieser says no one else monitors where the tours go. "Getting a boat out is expensive for researchers," he notes. "If we're already out, there's got to be something we can do to help." The company shares its information with the National Oceanic and Atmospheric Administration, which can make warnings as necessary.

The tour also observes humpback whale populations in the channel, taking observations about the number of whales present, whether there is a mother and calf present, what type of feeding activities the whales are taking, and how many boats are around the whales.

On land, the tour visits Mendenhall Glacier, where it samples water in Steep Creek, for benthic macroinvertebrates—small, spineless bugs that are found in healthy streams with clean water. These insects are found above the road that crosses the creek, but not below, and the Alaska Department of Environmental Conservation is investigating to see if the reason is related to runoff from the road or a natural cause such as turbidity or lack of sediment.

No particular skills or knowledge are required for the citizen science tour, and Gieser says that families with children are regular participants. "As long as they're interested in science, they have a good time," Gieser says.

WETLANDS AND WATERWAYS
—

AGE RANGE All ages **TIME FRAME** May through September
REGION Juneau, Alaska **FEE** $195 for adults; $99 for children

Great Lakes Marsh Monitoring Program

www.birdscanada.org/volunteer/glmmp

Delivered binationally since 1995 by Bird Studies Canada in partnership with Environment Canada and the US Environmental Protection Agency, the Great Lakes Marsh Monitoring Program monitors marsh birds and amphibians in the Great Lakes basin in Canada and the United States.

"When we started, there was a complete lack of information of what was going on in the wetlands around the Great Lakes," says Volunteer Coordinator Kathy Jones. Now the project has built a sizable database, and many of the marsh birds in it are not monitored by other programs. Data is used in scientific publications, shared with government and local partners for conservation decisions, presented at the State of the Lakes Ecosystem Conference every other year, and shared with volunteers through annual newsletters and formal reports every five years.

Amphibians and several birds, particularly the ones most dependent on marshes, have shown population declines over the course of the survey, while generalist bird species have experienced population increases.

For participants, "it's an excuse to go out in nature and do something positive," says Jones. "When they look at the marsh, they start taking ownership of it."

The amphibian and bird surveys are separate, although volunteers are welcome to contribute to both. To collect data about amphibians, MMP volunteers perform three surveys per year between March and July 5. Surveys take place on routes that volunteers choose in accordance with the program's standards. Some routes are surveyed during morning hours, while others are surveyed in the evenings before dark. At several monitoring stations per route, monitors listen for three minutes to identify the number of frog and toad calls and the species present.

Marsh bird surveyors conducting two surveys per year between mid-May and July. Some participants are assigned specific routes, while others determine their own based on the program's standards. At each station, they record all individuals of nine focal species seen or heard, all other bird species seen within one hundred meters, and all aerial foragers or birds seen flying. Participants observe for fifteen minutes at each station, divided into three parts: a five-minute silent listening period, a five-minute period where they play a CD intended to attract the calls of normally secretive species, and another five minutes of silence.

Jones says that the bird survey requires intermediate skills, including the ability to identify about fifty species by sight and sound. The amphibian survey requires less knowledge, which can generally be acquired through the project's training handbooks and CD, although most routes require strong mobility and comfort in a wetland at night.

Because much of the survey is based on sound, Jones says it's not necessarily suitable for families or other groups. But participants do include students looking for experience in their field. "For interested teens, it can be an entry into organized science."

AGE RANGE Teens and up **TIME FRAME** March through July **REGION** The Great Lakes in both US and Canada **$ FEE** None

---

Hudson River American Eel Research Project

www.dec.ny.gov/lands/49580.html

The New York Department of Environmental Conservation sponsors a research program in which teams of scientists, students, and citizen scientists collect information about the American eel in the Hudson River.

As a juvenile, the American eel migrates more than one thousand miles from the Sargasso Sea in the middle of the Atlantic Ocean, arriving in the Hudson as a two-inch-long glass eels. Project participants make observations at least one time per week in April and May, checking a net designed to catch migrating fish and recording information about the number and size of eels and environmental and tidal data.

American eel populations have declined in many eastern rivers for unknown reasons. The project hopes to investigate the reason. Data are also used in state management plans.

WETLANDS AND WATERWAYS —

AGE RANGE All ages **TIME FRAME** April and May
REGION Hudson River, New York **$ FEE** None

PRIDE Shorebird Surveys   www.lapurisimaaudubon.org/PRIDE.html

The La Purisima Audubon Society and PRBO Conservation Science host the Proud Residents Investing in a Diverse Estuary Shorebird Survey at the Lower Santa Ynez River near Vandenberg Air Force Base, California.

"Documented studies of the Lower Santa Ynez River are somewhat rare. It is nice to now have an established cadre of interested parties providing diverse and invaluable input into this project," says Tamarah Taaffe, treasurer of the La Purisima Audubon Society.

The river is an important bird area, because heavy storms can expose mudflats that provide habitat for shorebird populations. A bridge that was constructed in the early 1940s reduced mudflat habitat, and while the bridge washed out in 1969, remaining berms prevent those mudflats from reforming. The project is studying whether it's feasible to remove the bridge remnants, and to document conditions beforehand to assess the impact if that comes to pass.

Participants of all experience and knowledge levels are welcome. Surveys are conducted by at least four people per group, so new birders can be teamed with those with more experience. While observation does require quiet, Taaffe says that a family with children is among the regular participation.

Surveys last five to six hours, although beginners can join partway through. They consist of taking counts of shorebird species at several points along a set route every other week. The full survey involves hiking through marshy terrain, although one of the observation points is accessible to disabled volunteers.

Taaffe says that volunteers have a wide range of motives for their involvement. Some are students receiving credit, while others were preparing for internships and practicing skills in field work, and still others simply enjoyed their time on the river. "The main reward is gaining a heightened awareness of our estuary and its inhabitants—how nature and wildlife changes with the seasons," Taaffe adds.

AGE RANGE Preteen and up　**TIME FRAME** August through May
REGION Lower Santa Ynez River, California　**FEE** None

Upper Merrimack Monitoring Program

www.merrimackriver.org/programs/ummp.php

Created by the Upper Merrimack River Local Advisory Committee in 1995, the Upper Merrimack Monitoring Program conducts water quality testing at seventeen sites on the Merrimack River to give an indication of the river's health. The information volunteers gather helps to inform local, state, and high-level government decisions about which bodies of water will be added to the state 303(d) list for reporting as impaired under the Clean Water Act. Industry and academia also use and cite the data and other information produced by the Program

"People feel like they're not so helpless," says Michele L. Tremblay, director of the program. "They can see the health of the river in an up-to-date way."

Participants collect water samples every other week for eight to ten weeks. They also place rock baskets on the river bottom at each site to collect macroinvertebrates for annual testing. Those animals are identified at Bug Nights, a gathering that happens several times each month over the winter months to examine the macroinvertebrates and what they indicate about the river health. Bug Nights are both research-oriented and social, attracting a wide audience, from kids to retirees.

Tremblay says that volunteers come to the program for a number of reasons. The social aspect is a draw for some, while homeschoolers can participate as part of their curriculum, and retirees can stay active while making a contribution. "For people who went to school for sciences but can't get a job in their field, it's one of the only ways to use their skills," she says.

The project also includes an educational component. Representatives frequently present at schools, libraries, or civic organizations or give hands-on demonstrations at sites on the river.

AGE RANGE All ages   **TIME FRAME** August through April
REGION Upper Merrimack River Corridor, New Hampshire
FEE None

Vermont Vernal Pool Mapping Project

www.vtecostudies.org/VPMP

Vernal pools are temporary ponds that are dry for part of the year but fill up from winter rains or snow melt. These conditions provide an ideal habitat for certain plants and animals, particularly amphibians and insects that would be threatened by fish in a permanent pond. But because of their temporary nature, vernal pools aren't systematically mapped like other wetlands. "Therefore we don't know where they are on the landscape, and cannot effectively protect them from development or disturbance as we can with larger, mapped wetlands," says Steve Faccio, conservation biologist for the Vermont Center for Ecostudies.

The Vermont Vernal Pool Mapping Project is the center's citizen science effort to find and map these habitats throughout the state to improve conservation planning in the state. The project has already identified potential vernal pool sites through the use of aerial photography. Citizen scientists visit those sites to confirm whether the vernal pool exists and to collect information about the physical characteristics of the pool and what animals can be found there. They also record precise coordinates using a GPS. The data is used by state and local offices and conservation commissions to make land-use decisions.

Visiting potential vernal pool sites and finding animals there is a task well suited for families. "In fact, 8-year-olds are probably the best frog and salamander finders around," Faccio notes.

AGE RANGE All ages   **TIME FRAME** Primarily spring   **REGION** Vermont
FEE None

Water Action Volunteers
Citizen Stream Monitoring

http://watermonitoring.uwex.edu/wav

Water Action Volunteers' Citizen Stream Monitoring program encourages Wisconsin citizens to improve the water quality of the state's streams and rivers through regular monitoring. Project goals include education of Wisconsin citizens and advocacy for water quality as well as gathering information about stream health to inform management decisions.

"We're building a population of people who care about the natural environment and want to care for it in the future," says Kris Stepenuck, volunteer stream monitoring program coordinator.

There are three levels to the project. Basic monitoring has no prerequisites. Participants monitor four indicators of stream health—oxygen levels, temperature, transparency, stream flow—monthly from April through October at a stream or river of their choice. They also take annual observations of habitat conditions and semi-annual observations of macroinvertebrate life. Site visits generally take one to one and a half hours.

A second level seeks to gather more detailed information about waterway status and trends, and participants need some monitoring experience either at the first level or in another program. Participants receive several hours of training to learn the same monitoring methods used by Department of Natural Resources staff, including using handheld meters to assess dissolved oxygen and pH, checking transparency, and maintaining waterway monitors that give hourly temperature readings. The third level, which also requires prior monitoring experience, comprises special research projects, such as collecting water samples for phosphorous monitoring or conducting bacterial surveys.

Data that volunteers collect are publicly available in an online database. Monitors can also use the data they collect to have local impacts. Stepenuck says one monitor found low levels of dissolved oxygen in water caused by a company's illegal discharges. That volunteer reported the situation to the DNR, which rectified it.

"Surveys say participants are more active in their communities and share their knowledge with others," Stepenuck says. Many monitor in teams, which affords the opportunity to build friendships. "One team said, 'Our favorite part of monitoring is going out for a beer at the end of the day,'" she notes.

Family groups are welcome as well. "Some families live near a waterway, and monitoring is their family activity," Stepenuck says. In other cases, grandparents share the monitoring activity with their grandchildren. She hopes that this can build a long-term commitment to the environment in children. "I was involved [in monitoring programs] when I was a teen," Stepenuck says. "It helped build a sense of helping in the local community. I like to think that's happening for kids involved today."

AGE RANGE All ages **TIME FRAME** April through October, with some winter monitoring in urban areas **REGION** Wisconsin **$FEE** None

Resources

ADULT BOOKS

A Field Guide to the Animals of Vernal Pools by Leo Kenney and Matthew Burne (Massachusetts Division of Fisheries and Wildlife, 2001).

Flotsametrics and the Floating World: How One Man's Obsession with Runaway Sneakers and Rubber Ducks Revolutionized Ocean Science by Curtis Ebbesmeyer and Eric Scigliano (Smithsonian, 2009).

Marshes: The Disappearing Edens by William Burt (Yale University Press, 2007).

Moby-Duck: The True Story of 28,000 Bath Toys Lost at Sea and of the Beachcombers, Oceanographers, Environmentalists, and Fools, Including the Author, Who Went in Search of Them by Donovan Hohn (Penguin, 2012).

Ocean: The World's Last Wilderness Revealed by Robert Dinwiddie (DK Publishing, 2006).

Salt Marshes: A Natural and Unnatural History by Judith Weis and Carol Butler (Rutgers University Press, 2009).

Swampwalker's Journal: A Wetlands Year by David M. Carroll (Mariner, 2001).

Wetlands: Environmental Issues, Global Perspectives by James Fargo Balliett (M. E. Sharpe, 2010).

CHILDREN'S BOOKS

About Habitats: Wetlands by Cathryn Sill (Peachtree, 2008). Ages 6–8.

Looking Closely at the Shore by Frank Serafini (Kids Can, 2008). Ages 4–7.

Marshes and Swamps by Gail Gibbons (Holiday House, 1999). Ages 6–8.

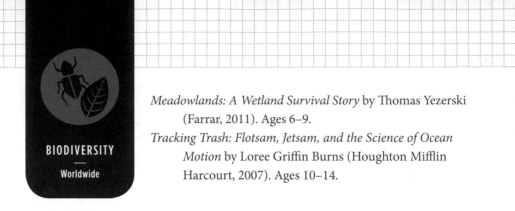

Meadowlands: A Wetland Survival Story by Thomas Yezerski (Farrar, 2011). Ages 6–9.

Tracking Trash: Flotsam, Jetsam, and the Science of Ocean Motion by Loree Griffin Burns (Houghton Mifflin Harcourt, 2007). Ages 10–14.

Biodiversity for Beginners

Many citizen science projects are ambitious efforts to catalog all life in a certain area—or even the whole world. From a participant's standpoint, they offer an excellent opportunity to learn about a wide range of creatures, and often plants as well. These are often good programs for beginners to explore general interests.

WORLDWIDE

All Taxa Biodiversity Inventory   www.dlia.org

An All Taxa Biodiversity Inventory seeks to document and identify all plants and animals in a specific area. Discover Life in America, which coordinated the All Taxa Biodiversity Inventory (ATBI) in the Great Smoky Mountains National Park in Tennessee, has documented more than seven thousand species never identified in the park before, and more than nine hundred of those were new to science.

"It's hard to conserve or protect things you don't know about," says Todd Witcher, executive director of Discover Life in America. "The park is full of things that haven't been documented or studied."

While individuals can contribute their observations to Discover Life's database, most of the information in it comes from scientist-led projects. For example, one scientist led research into arthropods that are associated with trees in the park whose numbers are declining. Citizen volunteers were trained to collect and sort mites, flies, and beetles in order to identify and determine what animals were affiliated with what trees.

Each ATBI is independent. In Tennessee, Discover Life in America offers regular training sessions so prospective participants can join projects when they arise.

Citizen science is "the cutting edge of how we're studying the park," Witcher says. Beyond the specific research projects, the information helps inform how the park is managed and developed. Many species are endemic to the park and found nowhere else. Among other benefits, knowing where they are ensures that trails, campsites, parking lots, and buildings can be built without inadvertently damaging the population.

BIODIVERSITY
—
Worldwide

Simply making this kind of observation and learning about the park's biodiversity is inherently valuable for participants. "Just driving into the park and seeing a bear is missing the point," Witcher says. "Biodiversity is a difficult term to understand, but getting citizens involved helps them to understand."

AGE RANGE All ages   **TIME FRAME** Year-round   **REGION** Many locations across the country   **FEE** None

National Geographic BioBlitz

www.nationalgeographic.com/explorers/projects/bioblitz

A BioBlitz is a twenty-four-hour inventory of all living things in a specific area. In honor of the US National Park Service's centennial in 2016, the National Geographic Society and the National Park Service are conducting an annual BioBlitz in a new national park each year leading up to the centennial.

The 2011 BioBlitz took place in Saguaro National Park near Tucson, Arizona. Five thousand participants explored the park and discovered more than four hundred species that had never been recorded in the park before. Prior BioBlitzes took place at Rock Creek Park in Washington, D.C., the Santa Monica Mountains National Recreation Area in California, Indiana Dunes National Lakeshore, and Biscayne Bay, Florida. The 2012 BioBlitz was held at Rocky Mountain National Park in Colorado.

Participants can join inventory teams to have the opportunity to do fieldwork with scientists and "highlight the importance of protecting the biodiversity of these extraordinary places and beyond," the BioBlitz website says. Advance registration is recommended, and participating in an inventory team requires some walking and about two and a half hours.

For those who can't make it to the *National Geographic* BioBlitz, there are local options. "A BioBlitz can happen in most any geography—urban, rural, or suburban—in as large an area as a national park or small as a schoolyard,"

the project's website says. Project Noah (projectnoah .org) is a website created by New York University and supported by *National Geographic* to document the world's organisms. It is suitable for finding or organizing BioBlitzes locally or based on specific interests. (See Project NOAH, page 139.)

AGE RANGE Recommended for 8 and up   **TIME FRAME** Late summer or fall; check website for specific dates   **REGION** Check website for future sites **$ FEE** None

The GLOBE Program   www.globe.gov

The GLOBE (Global Learning and Observations to Benefit the Environment) program is an international, hands-on science and education program that connects a network of students, teachers, and scientists to better understand, sustain, and improve the earth's environment at local, regional, and global scales.

GLOBE involves students of all ages in scientific discovery of the earth as a system through activities and scientific data collection. Students connect with their community and with other students and scientists worldwide who are contributing to global studies of the environment through the sharing of data and science projects. Additionally, the GLOBE Program provides training and professional development for all teacher participants. With GLOBE materials, teachers learn how to spark their students' interest in earth system science and not only learn about science, but actually participate in it.

"If you can connect students and teachers with scientists, it piques their interest and queues up the next generation of scientists and critical thinkers," says Ron Zwerin, GLOBE communications manager. "From a STEM [science, technology, engineering, and math] perspective, if we don't get to these students by fourth grade, we're going to lose them."

Participation in GLOBE is open to teachers in any of the participating countries who become trained in GLOBE scientific protocols. These teachers learn how to mentor students in taking, reporting, and using high-quality data and using these data to perform investigations.

"Students can gain extensive scientific, research, and critical thinking skills," Zwerin says. "The project also helps them see local information from a global standpoint."

Parents interested in bringing GLOBE into their child's school can find information on the project's website.

⚥ AGE RANGE School-age children **▦ TIME FRAME** Year-round
⊕ REGION Worldwide **$ FEE** None, although some instruments may need to be purchased

iNaturalist

www.inaturalist.org

At iNaturalist, people can share their observations of the plants and animals they see in nature and maintain life lists of their discoveries. After creating a free account, site visitors can upload photos and details about when and where they made their observations.

The site was created in 2008 by Nate Agrin, Jessica Kline, and Ken-ichi Ueda as a master's final project at the University of California at Berkeley School of Information. It began when Ueda realized that many people were like him and kept records of their explorations of nature. "I thought it would

At iNaturalist (www.inaturalist.org), people can share their observations of the plants and animals they see in nature and maintain life lists of their discoveries.

be cool to have a website to share and store data in one place," he says.

Data from the site goes into the Global Biodiversity Information Faculty, an international database of nature records used by researchers. The site also hosts dozens of citizen science inventories. Some focus on specific regions, while others seek to catalog specific classes of organisms—or even individual species. Observations can be contributed to multiple projects on the site. Site users can also create their own projects based on their own interests or research goals.

The site has an active community that can help to identify photos, which makes it a useful educational tool. "Every time you learn the name of an organism, it's like learning the name of a person," Ueda says. " Learning the names of different organisms helps you tell them apart, and once you can do that, you begin to think about how they relate to one another. Naming is a gateway into ecological curiosity."

Those identifications help to ensure the scientific value of the observations. Scott Loarie, iNaturalist codirector, says that many scientists welcome observations from amateurs if their quality can be verified. As a result, the site gives anyone who sees something rare or noteworthy an avenue to get their observation into the scientific community, even if they don't realize that their observation is unusual.

Unusual but accurate observations can raise interesting questions. One high school student reported a collared lizard in Sonoma County, California, four hundred kilometers outside of its normal range. That gave researchers enough information to investigate whether the lizard was simply a pet that had been released or if it was part of a previously unknown population.

Some unusual observations turn out to be errors, but even those have value. For example, if someone reports a tiger in Iowa, that observation is clearly incorrect. "But if a person never adds a tiger in Iowa, there's no start to the conversation," Ueda says. That "tiger" could actually be a bobcat or lynx, and when corrected, it provides useful information to researchers and helps the original observer to learn.

See also Global Amphibian BioBlitz / Global Reptile BioBlitz (page 55), Hawaii Sea Turtle Monitoring (page 63), and Summit County Citizen Science Inventory (page 142), which are all hosted on iNaturalist.

AGE RANGE All ages **TIME FRAME** Year-round **REGION** Worldwide **FEE** None

NatureMapping   www.naturemappingfoundation.org

BIODIVERSITY
—
Worldwide

The NatureMapping Foundation seeks to collect data and monitor wildlife through casual hikes, bioblitzes, and longer-term monitoring projects.

"Decisions are being made about land management, zoning, habitat modifications, and now climate change with little to no data," says Karen Dvornich, NatureMapping national director. The information that participants collect can be used to inform those decisions at the local, state, and regional levels.

Participating in a NatureMapping project affords the opportunity to take action to benefit the environment. "Participants can learn what lives with them in their own communities and then make a choice what they as individuals can do to restore, maintain, or improve their habitats," Dvornich says. Projects are also often conducted in groups, so they offer a chance to meet and learn from other interested people.

While individual NatureMapping projects vary, Dvornich says that many are suitable for all ages and that several elementary teachers have been trained to incorporate projects in their classes. "I've had teachers take our workshops because their students were bringing in bugs to be identified and they couldn't and they realized it would be fun to add to their curriculum," she says. "Kids love to learn something new."

A network of NatureMapping centers across the country facilitates local projects. There are NatureMapping centers or coordinators in fifteen states. People interested in NatureMapping can contact their local center through the project website.

People without a nearby NatureMapping center can still use the program's ideas to monitor wildlife and compare seasonal changes. "You don't have to have a NatureMapping Center or program in your state to begin 'mapping nature,'" Dvornich says.

Ⓨ **AGE RANGE** All ages ▦ **TIME FRAME** Year-round ⊕ **REGION** Fifteen states with NatureMapping Centers; see website for details $ **FEE** None

Project NOAH   www.projectnoah.org

Project NOAH (Networked Organisms and Habitats) is an online platform supported in part by National Geographic Society where users can document

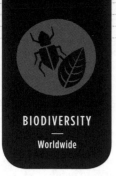

the wildlife and plants they see and contribute their observations to a variety of citizen science projects.

Participants can sign in through existing social media accounts such as Facebook or Twitter, or by creating student or teacher accounts. Then they upload photos of their observations, along with descriptions of the location, habitat, time and date seen, and identification, if known. "Part of the value proposition is that if you see something you don't know, someone in the community can identify it for you," says Yasser Ansari, Project NOAH cofounder. He adds that there is a significant focus on education, so corrections are made in a friendly manner.

Users can also assign their photos to one or more missions on the site. Missions collect observations in specific categories, some based on type of animal, others by habitat type or locality, and some by animal behaviors such as migrations or mimicking environments. Some align with existing citizen science projects—including Project Squirrel, the Lost Ladybug Project, and the National Geographic BioBlitz—although not all of them are formally connected. Others are just for fun. "Some of the most popular are ones I wouldn't have thought of," Ansari says, including one on animal architecture.

Some missions have a more important purpose, though. One, based in Colombia, documents wildlife that would be destroyed by a proposed dam development. Parents or teachers can create missions tailored for their children or students to explore their local environment.

Project NOAH is a worldwide network, so participants can connect with people well outside their neighborhood. Ansari tells of one teen in Malaysia who has discussed on the site's forums how he's not generally comfortable in social situations. But he does take and share outstanding nature photos, particularly of arthropods. Through the site, "he's talking and engaging with people on the other side of the planet who are interested in nature," Ansari says.

Project NOAH is open to anyone, although schools are particularly targeted. The site also offers Project NOAH for Teachers, where teachers can create student accounts and private classroom missions. That subsite includes a literacy component where teachers can require students to describe their observations before getting identification.

Project NOAH is available as a free mobile app.

AGE RANGE All ages   **TIME FRAME** Year-round   **REGION** Worldwide **$FEE** None

Wildlife Health Monitoring Network/ Wildlife Health Event Reporter

www.whmn.org/wher

The Wildlife Health Monitoring Network integrates a variety of data about wildlife disease patterns and their potential impact on the health of humans and domestic animals. The Wildlife Health Event Reporter allows citizen scientists to report sick, injured, or dead wild animals to the network. Natural resource managers, researchers, and public health officials can use the data to design and coordinate disease control strategies, detect common and emerging diseases and biosecurity concerns, and explore the connections between human and wildlife diseases.

"Seventy-five percent of recent emerging infectious diseases in humans began as animal infections, and most of these have involved wildlife," said US Geological Survey scientist and WHER developer Joshua Dein in a press release announcing the Reporter. "If these tools had been available ten years ago, we might have had an earlier identification of West Nile virus by people reporting that they were seeing dead crows in their backyards."

The Wildlife Health Event Reporter is an opportunistic citizen science project; participants can report whenever they choose and are not expected to make a minimum commitment. A mobile app is available.

AGE RANGE All ages   **TIME FRAME** Year-round   **REGION** Worldwide   **FEE** None

REGIONAL

Mount Rainier Citizen Science   http://rainiervolunteers.blogspot.com

Mount Rainier near Seattle offers a schedule of citizen science volunteer opportunities from July through September. Available projects are diverse in subject and cover butterflies, amphibians, vernal pools, wilderness soundscapes, and other topics.

Training is provided for all projects, so no scientific experience is required. Most projects are physically strenuous and often require hiking and camping in the park.

Volunteer opportunities are posted as they come up on the park's volunteer blog.

BIODIVERSITY
—
Regional

† **AGE RANGE** Varies ▦ **TIME FRAME** Year-round
⊕ **REGION** Mount Rainier, Washington $ **FEE** None

Summit County Citizen Science Inventory

www.inaturalist.org/projects/summit-county-citizen-science-inventory

Metro Parks, Serving Summit County, has a natural resources department that has been cataloging the biodiversity of the county's parks to help manage land for ten years.

But the park's staff, which usually includes four or five biologists, can't be everywhere, and they can't know everything. None of the biologists are experts in fungi, for example. "We know there's a lot more people out there looking at wildlife, and we would like to include their observations in our database," says park biologist Marlo Perdicas.

Metro Parks started the Summit County Citizen Science Inventory in 2012 to use citizens to help supplement the park's biodiversity monitoring.

"We are collecting nature data at an unprecedented rate," Perdicas says. The park had about ten thousand observations before starting the citizen science inventory. "In just a couple of months, we have already generated eight thousand observations," from volunteers, she adds, which include documentation of fungi and insects that nobody on staff was an expert in. "Those observations have been priceless," Perdicas says.

The project collects observations of all forms of life in the county, including animals, plants, and fungi. It is hosted on iNaturalist, and participation is open to anyone with a free account. Participants post photos of their observations with information about the time and location of those sightings.

Metro Parks has an extensive volunteer program with more than five hundred participants, and some of its volunteer projects include citizen science monitoring of stream quality, amphibians, and nesting birds. (See www.summit metroparks.org for details.) The Citizen Science Inventory is not limited to regular volunteers, and it allows residents to contribute without formally registering with the park.

No special skills or knowledge are required to contribute to the inventory. iNaturalist requires photos with all observations, which can then be identified by the community. It also encourages participation by maintaining life lists of species observed and providing a leaderboard to recognize those who contribute the most total observations and the most unique species in each project.

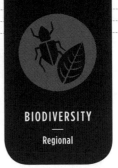

BIODIVERSITY

Regional

Ⓨ AGE RANGE All ages **🗓 TIME FRAME** Year-round ⊕
REGION Summit County and surrounding counties in northeast Ohio **$ FEE** None

Texas Nature Trackers

www.tpwd.state.tx.us/trackers

Sponsored by the Texas Parks and Wildlife Department, Texas Nature Trackers is a collection of citizen science monitoring efforts for box turtles, amphibians, prairie dogs, horned lizards, hummingbirds, mussels, bumblebees, and whooping cranes. Participants survey species of interest—several of which have experienced dramatic declines—on their property or on public land.

In general, the TNT surveys are opportunistic, and participants may report a species whenever they see it. Some surveys do require formal training, however. Mussel Watch volunteers, for example, need to learn to identify species and how not to impact mussel habitats while observing, because several mussel species are severely imperiled. Volunteers must be licensed to handle them.

But even Mussel Watch, the most involved of the programs, is accessible to all ages, says Marsha May of the Texas Parks and Wildlife Department. Particularly good for families is Amphibian Watch, where participants can learn to identify species online or at a local workshop and can make observations at any time.

Each individual TNT program has a biologist attached who records and manages the data, and all data collected by TNT programs is available to researchers and members of the public. It is used specifically in the state's conservation action plan, and data on rare species is added to the Texas Natural Diversity Database, which is consulted by the Texas Department of Transportation before construction projects. Volunteers also receive annual reports. "A lot of folks really like having knowledge that they are part of something bigger," May says.

Texas Nature Trackers began in the 1990s to help collect data requested by the US Fish and Wildlife Service. It continues expanding and focusing on species of concern. Whooper Watch, for example, is a new program to monitor

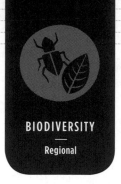

changes in wintering locations of whooping cranes possibly due to a major drought in 2011. "Cranes normally winter in coastal Texas, but last winter surprised us," May says. "Family groups ended up at Granger Lake," which is several hundred miles north of the crane's normal wintering grounds. The project seeks to determine whether the drought is the reason, and if not, what might be.

AGE RANGE All ages　**TIME FRAME** Year-round　**REGION** Texas　**$ FEE** Varies; some projects require a small fee (up to $10) to cover the cost of training or monitoring materials

Urban Ecology Center

http://urbanecologycenter.org/what-we-do/citizen-science.html

The Urban Ecology Center is a nature center in Milwaukee that uses several city parks for research that is driven by citizen scientists. Volunteers research and monitor local birds, bats, snakes, small mammals, frogs, turtles, invertebrates, monarchs, and vegetation.

"We work with large-scale projects, but we're focused on local issues," says Tim Vargo, manager of research and citizen science. For example, volunteers

Volunteers banding birds for the Urban Ecology Center.

Photo by Corinne Palmer

have been working with professional scientists to collect information about a locally threatened snake. A main goal is to have the data citizen scientists collect about the snake be used by policy makers in conservation plans.

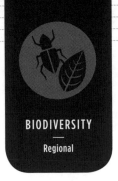

About 150 people participate in the center's citizen science projects. Prospective participants apply for projects they are interested in through the center's volunteer coordinator. See the website for details.

AGE RANGE Teens and up **TIME FRAME** Year-round **REGION** Milwaukee, Wisconsin **$ FEE** None

Resources

ADULT BOOKS

Calls beyond Our Hearing: Unlocking the Secrets of Animal Voices by Holly Menino (St. Martins, 2012).

The Exultant Ark: A Pictorial Tour of Animal Pleasure by Jonathan Balcombe (University of California Press, 2011).

Last Child in the Woods: Saving Our Children from Nature-Deficit Disorder by Richard Louv (Algonquin, 2008).

Super Species: The Creatures that Will Dominate the Planet by Garry Hamilton (Firefly, 2010).

CHILDREN'S BOOKS

Animal Homes by Angela Wilkes (Kingfisher, 2003). Ages 5–7.

Castles, Caves, and Honeycombs by Linda Ashman (Harcourt, 2001). Ages 4–6.

Citizen Scientists: Be a Part of Scientific Discovery from Your Own Backyard by Loree Griffin Burns (Holt, 2012). Ages 8–12.

Eco-tracking: On the Trail of Habitat Change by Daniel Shaw (University of New Mexico, 2011). Ages 12–14.

The Kingfisher Encyclopedia of Life (Kingfisher, 2012). Ages 8–12.

Life on Earth—and Beyond: An Astrobiologist's Quest by Pamela Turner (Charlesbridge, 2008). Ages 10–15.

Nature in the Neighborhood by Gordon Morrison (Sandpiper, 2008). Ages 6–9.

Wild Tracks! A Guide to Nature's Footprints by Jim Arnosky (Sterling, 2008). Ages 6–8.

Outer Space

Astronomy is one field of science that has a long history of amateur contributions. Given the vastness of space, that's hardly surprising. No matter how much effort professional astronomers put in, there's no chance they can monitor even a fraction of the skies. These projects offer opportunities for citizen scientists to focus their efforts.

See also Zooniverse, page 50, under Analytical Games and Puzzles.

American Association of Variable Star Observers

www.aavso.org

AAVSO is an organization of amateur and professional astronomers, rather than a specific citizen science project, but it collects and coordinates data much like any of the projects in this directory would.

Variable stars are stars that change brightness, whether due to changes in the star itself or other factors that change the amount of light from the star that reaches earth. "Research on variable stars is important because it provides information about stellar properties, such as mass, radius, luminosity, temperature, internal and external structure, composition, and evolution," the AAVSO website says. "Some of this information would be difficult or impossible to obtain any other way."

The organization maintains the AAVSO International Database, a collection of more than twenty million variable star brightness estimates dating back more than one hundred years. Amateur and professional astronomers can submit observations to the database, where it will be included after AAVSO validation. This database is a major contributor to scientific research: dozens of papers per year cite the information it contains.

AAVSO also maintains the International Variable Star Index, a repository of known variable stars. Astronomers who believe they have found a new variable star can check against the index before reporting it.

From 2009 to 2011, AAVSO operated Citizen Sky, a citizen science project to study epsilon Aurigae, a bright star in the Auriga constellation that astronomers have been studying for more than 175 years. Epsilon Aurigae is a binary star, which means there are actually two stars orbiting around each other. But while one star is quite bright, the other emits much less light than expected for a star of its size. Citizen Sky examined the star while the secondary star eclipsed the primary, a phenomenon that happens only every 27.1 years.

OUTER SPACE
—

AGE RANGE All ages **TIME FRAME** Year-round **REGION** Worldwide **FEE** $60 for an annual membership

American Meteor Society Visual Observing Program

www.amsmeteors.org/ams-programs/visual-observing

The American Meteor Society collects data from amateur and professional astronomers about meteors, meteor showers, fireballs, and related phenomena. Its Visual Observing Program encourages amateurs to provide the data they collect about brief meteor outbursts and fireballs that might not be observed by any other source.

"Meteor Science is one of the few remaining astronomical fields where amateur astronomers, equipped with only their eyes, can provide a valuable service to the planetary science community," the project's website says. Participants collect details about each meteor they see, including its brightness, color, type, and speed. Data contributed to the project is shared with interested organizations and professional and amateur astronomers for use in their research.

Meteors are best viewed during the early mornings. While there are peak viewing days in August, October, December, and January each year, meteors can be viewed every night of the year. The program encourages monitoring of those "minor showers" among motivated observers.

AGE RANGE All ages **TIME FRAME** Year-round **REGION** Worldwide **FEE** None

Galaxy Zoo

www.galaxyzoo.org

Galaxy Zoo is a Zooniverse project that seeks to examine how galaxies form. Participants examine images of hundreds of thousands of galaxies to classify their shape and features. Humans can make this kind of judgment better than computers can.

"Our strategy is based on the remarkable fact that you can tell a lot about a galaxy just from its shape," the project's website says. "Find a system with spiral arms, for example, and normally—but critically not always—you'll know that you're looking at a rotating disk of stars, dust and gas with plenty of fuel for future star formation. Find one of the big balls of stars we call ellipticals, however, and you're probably looking at a more mature system, one which long ago finished forming stars. The galaxies' histories are also revealed; that elliptical is likely to be the product of a head-on collision between two smaller galaxies, and smaller scars such as warped disks, large bulges or long streams of stars bear testament to the complexity of these galaxies' lives."

After registering and reading a short tutorial, participants can begin classifying pictures. Each image is classified by many participants in order to improve accuracy by gaining consensus.

The project started in 2007, and in its first year participants classified a million images taken from the Sloan Digital Sky Survey. As of 2012, the project is in its fourth incarnation, now analyzing ultradeep images of the universe from the Hubble Space Telescope's CANDELS survey. More than 250,000 people have participated so far.

Among Galaxy Zoo's unique accomplishments is the discovery of the Voorwerp. A Dutch volunteer found a blue splotch in one of the images he was classifying and posted to the project's forums to see if anyone knew what it was. Nobody did, not even the professional astronomers, but now the unknown object (*Voorwerp* means "object" in Dutch) can be researched in more depth. "It's something that is unique to a project like the Zoo," the project's website says. "Computers will slowly get better at classifying galaxies, but looking at an image and asking 'what's that odd thing?' remains uniquely human."

Y AGE RANGE All ages ▦ **TIME FRAME** Year-round ⊕ **REGION** Worldwide $ **FEE** None

---

GLOBE at Night

www.globeatnight.org

GLOBE at Night uses data collected by citizen scientists to measure the impact of light pollution on the visibility of stars in the night sky. Participants compare the sky where they are to charts provided by GLOBE at Night to determine the faintest stars that can be seen, and contribute data through the project's website.

"We're trying to get people to light more responsibly," says Connie Walker, creator of GLOBE at Night. That doesn't mean simply turning off street and building lights, she adds. Instead, the project hopes to persuade the public and policy makers to consider the impact of their artificial lighting.

In 2011, *National Geographic* held its BioBlitz, a twenty-four-hour count of animal species, in Saguaro National Park near Tucson. GLOBE at Night joined in, measuring star visibility at various points from the center of the city to the edge of the park. Through the park, participants found a difference of five orders of magnitude, and college students conducted studies and found that areas with darker skies had greater populations of endangered animal species.

GLOBE at Night has collected data since 2006. That information has shown a general brightening of the night skies. In 2006, 55 percent of participants reported seeing stars of magnitude 4 or greater (dimmer stars have higher magnitude ratings); in 2011, only 45 percent did. High schools and amateur astronomers have used this information to make maps to try to influence local lighting ordinances.

AGE RANGE All ages **TIME FRAME** January through May; check website for specific times **REGION** Worldwide **$ FEE** None

---

Lowell Amateur Research Initiative

www.lowell.edu/LARI_welcome.php

The Lowell Observatory in Flagstaff, Arizona, has a number of research projects that amateur astronomers can contribute to. "These projects span a broad range of technical skills and knowledge from taking very deep images of galaxies to monitoring small stars for transient events to data mining," the observatory's website says.

"Participants take great pleasure in discovering the wonders of the universe," says Bruce Koehn, LARI administrator. "That statement is just poetic language describing the desire of many folks to carefully analyze natural phenomena in an attempt to discover the way things work."

While the project is young, participants have already photographed dwarf galaxies, obtained light curves to demonstrate the rapid rotation of a few

young stars, and reported the positions of objects in the Kuiper belt outside Neptune, says Koehn. In the first round of projects, fifty-seven amateurs have worked on seven projects, although new projects will be created as necessary and projects will end as their research is completed.

Prospective participants must create an account on the Lowell Amateur Research Initiative website. After that, they create a profile that includes information about their astronomical interests, research experience, equipment and software they are proficient in using, location, amount of time they can contribute, and the research projects they are interested in joining. New profiles are evaluated every three months to match amateurs with projects that fit their interests and abilities.

Many, though not all, of LARI's research projects require quality equipment and the commitment to make observations whenever conditions make those observations possible, rather than when convenient. Most are aimed at serious amateur astronomers. "We generally expect the research associate to be capable of understanding the required mathematics, software, the scientific method, and so on," Koehn says. As a result, they are not necessarily suitable for all families, although they would be accessible to advanced high schoolers.

AGE RANGE High school and up   **TIME FRAME** Year-round   **REGION** Worldwide
FEE None

Moon Mappers   http://cosmoquest.org/mappers/moon

Moon Mappers is a project of CosmoQuest to analyze photos of the moon taken by the Lunar Reconnaissance Orbiter launched by NASA in 2009.

"Craters can tell planetary scientists a lot about a surface, such as its age, how thick the regolith [moon 'soil'] is, what kinds of erosion processes may have occurred, and what kind of material may be just under the surface," the project's website says. "To study all these things, we need to know where impact craters are, how many they are, and different things about them."

After registering and completing a tutorial, participants are shown a lunar surface image and asked to mark all craters of a minimum size. They may also be given an image with craters identified by computer to check, since computer algorithms can't identify features such as craters as well as humans can.

⍾ AGE RANGE All ages **▦ TIME FRAME** Year-round
⊕ REGION Worldwide **$ FEE** None

Stardust@home

http://stardustathome.ssl.berkeley.edu

Stardust@home is an online search for interstellar dust. The *Stardust* spacecraft captured samples from the comet Wild 2 in 2004.

But much of what the spacecraft collected was not interstellar dust. Original estimates were that about forty-five interstellar dust particles would be collected, and only four have been found so far.

Because the interstellar dust particles are so small—only a micron in size—researchers would need to make 1.6 million microscope observations in order to examine the entire sample. That would take the research team more than twenty years.

Instead, the project is using an automatic scanning microscope to create images of the collector for anyone to examine online. Participants are given a stack of images, called a *focus movie*, that show one small section of the collector both above and below its surface. Volunteers look for tracks that indicate where a dust grain might have entered the collect.

Prospective volunteers must take a short online tutorial and pass a track identification quiz before they are allowed to register. Once registered, participants can examine as many focus movies as they wish.

Each focus movie is shown to several volunteers to ensure that tracks are not missed. When several participants agree on a track, the project's researchers investigate it more closely to determine if it is in fact a particle of interstellar dust.

The project has a game element: volunteers are scored on accuracy, and a leaderboard tracks the top participants. Those who find a track that turns out to be made by interstellar dust will be credited as a coauthor on any paper announcing its discovery, and also have the right to give the particle its name.

⍾ AGE RANGE All ages **▦ TIME FRAME** Year-round **⊕ REGION** Worldwide **$ FEE** None

Resources

ADULT BOOKS

Confessions of an Alien Hunter: A Scientist's Search for Extraterrestrial Intelligence by Seth Shostak (National Geographic, 2009).

The Cosmic Connection: How Astronomical Events Impact Life on Earth by Jeff Kanipe (Prometheus, 2009).

The 50 Best Sights in Astronomy and How to See Them: Observing Eclipses, Bright Comets, Meteor Showers, and Other Celestial Wonders by Fred Schaaf (Wiley, 2007).

Gravity's Engines: How Bubble-Blowing Black Holes Rule Galaxies, Stars, and Life in the Cosmos by Caleb Scharf (Farrar, Straus and Giroux, 2012).

The Sun's Heartbeat and Other Stories from the Life of the Star that Powers Our Planet by Bob Berman (Back Bay, 2012).

CHILDREN'S BOOKS

Boy, Were We Wrong about the Solar System! By Kathleen Kudlinski (Dutton, 2008). Ages 6–8.

Cosmic! The Ultimate 3-D Guide to the Universe by Giles Sparrow (DK, 2008). Ages 10–14.

Exploring the Solar System: A History with 22 Activities by Mary Kay Carson (Chicago Review Press, 2008). Ages 10–14.

The Mighty Mars Rovers by Elizabeth Rusch (Houghton, 2012). Ages 11–15.

The Mysterious Universe: Supernovae, Dark Energy, and Black Holes by Ellen Jackson, photographs by Nic Bishop (Houghton, 2008). Ages 11–14.

Older than the Stars by Karen Fox (Charlesbridge, 2010). Ages 5–7.

Once upon a Starry Night by Jacqueline Mitton (National Geographic, 2009). Ages 6–9.

Stars by Ker Than (Children's Press, 2010). Ages 7–10.

13 Planets: The Latest View of the Solar System by David A. Aguilar (National Geographic, 2011). Ages 8–10.

A Zoo in the Sky: A Book of Animal Constellations by Jacqueline Mitton (National Geographic, 1998). Ages 6–9.

Plants and Fungi

Many plants are of significant conservation concern. Others are invasive plants that threaten to take root in new places and choke out the native plants. Both are of significant interest, and the following citizen science projects seek to monitor and evaluate them to improve natural habitats. Fungi, which actually falls into a kingdom of its own, is another area where citizen science projects have begun to flourish.

Global Garlic Mustard Field Survey

www.garlicmustard.org

Part of the Global Invasions Network, the Global Garlic Mustard Field Survey is a research project investigating why the quantity of invasive species like garlic mustard varies at different spaces and time.

Participants find populations of garlic mustard and collect information about their locations, the size of those populations, and the development of individual plants. Participants also collect twenty inflorescences, the part of the plant that includes its stem and pods, for analysis by researchers.

The project operates in North America, where the plant is an invasive species, and Europe, where it is native. Conventional wisdom suggests that invasive species are larger and more effective at reproducing in their adopted habitats than their native ones, but that idea has not been well tested. Measuring populations in both North America and Europe will help to provide data that will fill that gap.

AGE RANGE High school and up **TIME FRAME** Year-round **REGION** North America and Europe **FEE** None

Invaders of Texas

www.texasinvasives.org/invaders

Invaders of Texas is a project of the Texasinvasives.org partnership to search for and report harmful invasive species such as zebra mussels, emerald ash borers, or giant hogweeds.

"Invasive species are really bad for native plants," says Jessica Strickland, invasive species program coordinator at the University of Texas at Austin's

Lady Bird Johnson Wildflower Center. "People see the impact," especially when they are involved with identifying where those invasives grow.

Many participants are part of satellite groups—local groups that have partnered with the program to organize special event days to collect information. Individuals can also receive training to identify invasive species and gather information when they choose. For certain high-priority animal species, the program welcomes observations from anyone.

The project coordinates with the City of Austin and the Audubon Society of Texas to remove invasive species, and it works with appropriate agencies to validate and address high-priority reports. It also submits data to the Early Detection and Distribution Mapping System, a University of Georgia database that provides early alert for the distribution of invasive plants. Information gathered can help local, state, and federal managers determine their priorities.

While many participants are affiliated with programs such as Texas Master Naturalists and Texas Master Gardeners, Strickland says Invaders of Texas is suitable for all ages. "You go out and see something that doesn't look right," she says. "Take a photo and upload it at home, and it pops up on a map. It's very interactive, and your name is attached to the observation."

Texasinvasives.org also has an Eradicator Calculator feature to publicize eradications and quantify the value of volunteer efforts.

AGE RANGE All ages   **TIME FRAME** Year-round   **REGION** Texas   **FEE** None

---

Mountain Watch   www.outdoors.org/conservation/mountainwatch/mtplant.cfm

The Appalachian Mountain Club hosts Mountain Watch, a citizen science program to monitor plants in the Appalachian region. Participants take phenology observations (data about when recurring events take place, such as the leaves changing color in autumn) to help investigate how different mountain plants respond to environmental cues and track the timing of flowering and fruiting in the long term.

"Plants in cold limited ecosystems, such as alpine and other mountain environments, may act as sensitive bioindicators of climate change," the project's website says. Data are shared with the USA National Phenology Network and the Appalachian Trail Mega-Transect Monitoring Project.

The project is designed for participants to contribute to while hiking. Participants do not need to make regular observations, although there are provisions for those who can make repeat observations on a single route.

AGE RANGE All ages **TIME FRAME** Spring and summer **REGION** Appalachian Mountains **FEE** None

Mushroom Observer

www.mushroomobserver.org

Launched in 2006, Mushroom Observer is a website that collects observations and photos of mushrooms and other fungi—more than one hundred thousand so far. The site welcomes data from both amateurs and professionals.

"By some estimates less that 5% of the world's species of fungi are known to science," the project's website says.

Users can upload observations, including photos, identification if available, and notes about habitat and physical properties. For unidentified species, the site's community typically proposes identifications, often within twenty-four hours.

AGE RANGE All ages **TIME FRAME** Year-round **REGION** Worldwide **FEE** None

NY-NJ Trail Conference Invasives Strike Force

http://trails.rutgers.edu

A partnership that grew out of Rutgers University and the New York–New Jersey Trail Conference in 2011, the Invasives Strike Force monitors invasive plants along hiking trails in New York and New Jersey. Nonnative plants introduced to the environment by human activity, animals, or wind, can displace native plant life and harm the animals that depend upon the native species.

The project seeks to collect information about invasive plants along the trails and identify areas where invasive species can be removed in such a way to prevent them from spreading into uninvaded communities.

Participants need to attend a one-day training class, several sessions of which are held each May and June, to learn to identify common invasive

plants, how to collect data, and how to use a GPS. After training, each volunteer is assigned a section of trail approximately two miles long, where they are responsible for identifying and mapping invasive plants.

AGE RANGE All ages **TIME FRAME** Summer
REGION Parks in New York and New Jersey **$ FEE** None

PlantWatch

www.plantwatch.ca

Part of the NatureWatch suite of Canadian citizen science programs, PlantWatch encourages citizen scientists to report flowering times for thirty-nine plant species. "Our goal is to encourage Canadians of all ages to get involved in helping scientists discover how, and more importantly why, our natural environment is changing," the project's website says.

Participants select individual plants to monitor. They make observations at least every few days to report the first bloom, midbloom, and for trees and lilacs, leafing dates.

All data collected, about nine thousand observations so far, are publicly available on the project's website. Several peer-reviewed research papers have been published based on the data, including studies on trends in spring flowering dates, how flowering dates and other phenology trends might link to ocean temperature, and the value of linking on-the-ground observations to those made remotely.

AGE RANGE All ages **TIME FRAME** Primarily spring **REGION** Canada **$ FEE** None

Plant Community Monitoring

www.habitatproject.org/opportunity/monitor.html

The Chicago Wilderness Habitat Project conducts periodic audits of habitats in the Chicagoland area. These audits consist of about fifty volunteers sent to randomly selected spots in Chicago habitats to collect details about the type of plant species growing there, and particularly invasive species, in order to give a snapshot of the quality of the habitat.

"It's really been useful," says Karen Glennemeier, Habitat Project conservation scientist, adding that data have led to increased resources for restoration.

The project has conducted audits of Chicagoland grasslands and woodlands so far, as well as an audit of Cook County as a whole.

AGE RANGE Adult   **TIME FRAME** Irregular intervals
REGION Northeastern Illinois   **FEE** None

Plants of Concern

www.plantsofconcern.org

Plants of Concern is a monitoring program for rare plants in the Chicago area coordinated by the Chicago Botanic Garden, based on a biodiversity recovery plan developed by Chicago Wilderness in the mid-1990s. The plan, developed with the aid of more than 250 conservation agencies, focused on the status of endangered species and gaps in knowledge about them.

"Monitoring had been done all along," says Susanne Masi, Plants of Concern coordinator. "But it had never been done in a systematic way across

Photo by Ellen Powell

Volunteers participate in the Plants of Concern Illinois Beach Foray.

the region." One of the plan's recommendations was to develop a volunteer monitoring network to address that inconsistency. That network, Plants of Concern, officially began in 2001.

Volunteers conduct surveys once per year of one or more plant species on a site, measuring the size of the population of the species, soil conditions, whether the plants are reproducing or if there are seedlings present, threats to the population such as invasive brush or unauthorized trails, nearby native plants, and management activities. In all, participants monitor more than 175 species, many of which are endangered or threatened. "These species are the most threatened and most precious in a way, and they need attention," Masi says. Helping them has wider advantages as well. "If you clear the brush for one plant, others will benefit," she says.

"A huge part of our goal is to alert land managers to problems we've seen," Masi says. Each landowner where the project surveys receives reports of what the surveyors find. Project findings have been used to protect rare plants by giving landowners information to change mowing regiments or prescribed burns, or erecting cages to protect plants against deer browse. Data are also contributed to the Illinois Natural Heritage Database and used by researchers, including one study that analyzed the soils, genetics, and population changes of rare grasslike species, and another that conducted genetic analysis of species that were not reproducing.

Participants need some plant knowledge, although not expert-level identification skills, Masi says. A one-day workshop helps new volunteers learn what they need to participate. They also must be enthusiastic about finding native plants, able to make careful observations, and capable of moderate activity that often includes hiking and bending in hot weather. Because of this—and the fact that many of the species surveyed are endangered and their exact locations must be kept confidential—Masi says the program isn't ideal for young children, although many teens and some preteens happily join their parents in the field.

Participants report that the project has changed their lives, Masi says, inspiring them to learn more about plants and gain a greater awareness of the connections in nature. "Many have a spiritual connection to the earth and really care about what happens to the environment," Masi adds. Others enjoy meeting people who share their interests.

The project also conducts forays at specific sites, where participants visit a site as a group to measure and count a specific species. "It's a real group effort, and it's a lot of fun," Masi says.

PLANTS AND FUNGI

⌖ AGE RANGE Primarily high school and up **📅 TIME FRAME** Spring and summer **🌐 REGION** Northeastern Illinois and northwestern Indiana **$ FEE** None

---

Project BudBurst

www.budburst.org

Founded in 2007 by the National Ecological Observatory Network and the Chicago Botanic Garden, Project BudBurst is a network of thousands of citizen scientists who monitor plants as the seasons change. Specifically, participants collect data on plant phenology—when various plants grow leaves, flower, and produce fruit each year. That information is used to show how individual plant species respond to changes in climate—an important matter because many animals' developmental cycles are tied to plants. "As the climate warms, plants may become out of sync with the insects that pollinate them," the project's website says. "If an insect is still a larva when the flowers blossom, for example, it will not be able to fly from flower to flower to transport pollen."

Project BudBurst participants make observations of plants in their yards or neighborhoods a few times per week. They can select from the dozens of plants that are included in the project, including wildflowers, herbs, grasses, evergreen trees and shrubs, deciduous trees and shrubs, and conifers. The project also accepts one-time reports for certain plants.

After observation, participants report online the plant, its location, and its stage of development. Reported data is freely available for use from the project's website, which contains a wealth of information on each of the included species, including a map of development nationwide. BudBurst Buddies (neoninc.org/budburst/buddies) is a companion project that helps young children to make observations.

⌖ AGE RANGE All ages **📅 TIME FRAME** Year-round **🌐 REGION** US **$ FEE** None

Resources

ADULT BOOKS

The Bizarre and Incredible World of Plants by Wolfgang Stuppy and Madeline Harley (Firefly, 2012).

A Field Guide to Wildflowers: Northeastern and North-central North America by Margaret McKenny and Roger Tory Peterson (Houghton Mifflin Harcourt, 1998).

Mushroom by Nicholas Money (Oxford University Press, 2011).

Newcomb's Wildflower Guide by Lawrence Newcomb and Gordon Morrison (Little, Brown, 1989).

Plants of the Chicago Region by Floyd Swink and Gerould Wilhelm (Indiana University Press, 2012).

Tallgrass Prairie Wildflowers by Doug Ladd and Frank Oberle (Falcon/Globe Pequot Press, 1995).

What a Plant Knows: A Field Guide to the Senses by Daniel Chamovitz (Scientific American/Farrar, Straus and Giroux, 2012).

CHILDREN'S BOOKS

Eyewitness Plant by David Burnie (DK, 2011). Ages 8–14.

Flip, Float, Fly: Seeds on the Move by JoAnn Macken (Holiday House, 2008). Ages 6–9.

Lively Plant Science Projects by Ann Benbow and Colin Mably, (Enslow, 2009). Ages 8–12.

Living Sunlight: How Plants Bring the Earth to Life by Molly Bang and Penny Chisholm (Scholastic, 2009). Ages 6–8.

New Plants: See in the Soil Patch by Emily Sohn and Erin Ash Sullivan (Norwood House, 2012). Ages 8–10.

Plant Secrets by Emily Goodman (Charlesbridge, 2009). Ages 5–7.

Winter Trees by Carole Gerber (Charlesbridge, 2008). Ages 4–9.

Weather and the Seasons

Phenology is the study of the timing of recurring natural events such as migrations or plant flowerings. It raises important questions about the effects of climate change. Many species' life cycles are keyed to temperature changes that vary as temperatures get warmer earlier. But other species' changes are based on changes in light, which remain constant. These species rely upon each other, though, so when one changes before the other, what will the impact on the populations be? Many phenology projects seek the answers to that question.

See also Project Budburst, page 159, under Plants and Fungi.

Community Collaborative Rain, Hail, and Snow Network

www.cocorahs.org

Founded in 1998 at the Colorado Climate Center at Colorado State University in response to a flood at Fort Collins a year prior, the Community Collaborative Rain, Hail, and Snow Network (CoCoRaHS) is a volunteer effort to track precipitation nationwide.

"Everyone has seen where it's raining across the street but not on you," says CoCoRaHS Education Coordinator Noah Newman. CoCoRaHS participants measure precipitation—or lack of precipitation—in their area from as many locations as possible, which supplements National Weather Service data.

The information helps local water managers plan for water supply and demand, and can help regional flood managers predict where flooding may occur. NASA also plans to use CoCoRaHS data to help calibrate the Global Precipitation Measurement Satellite. Individual users can track precipitation reports on the project's website for their own purposes as well.

"A lot of people do it for the sense of being part of the community," Newman says. "They know their data is used and feel a sense of accomplishment." He adds that many schools and families contribute to the project, including homeschoolers who use it as part of their lessons. And for all ages, there's value in the act of observing. "Observing is something that takes practice," he says. "It opens your eyes to how little precipitation we get, especially when there's a drought."

AGE RANGE All ages **TIME FRAME** Year-round **REGION** US and Manitoba; the project will soon expand into the rest of Canada **$ FEE** None

IceWatch

www.icewatch.ca

IceWatch is part of the NatureWatch suite of Canadian citizen science programs. Participants monitor freezing and thawing dates of freshwater lakes and rivers to help identify changes in the natural environment.

"By analyzing citizen records, scientists have found that the freeze-thaw cycles of Northern water bodies are changing," the project's website says. "However, since climate change is not consistent across the country and there are large gaps in the current monitoring network, scientists require critical data from many more regions."

"Having more observers relative to what's changing in the natural world gives more information to detect changes," adds Marlene Doyle, NatureWatch manager.

Participants can select almost any lake or river, although those controlled by dams or those that run parallel to prevailing winds are less than ideal because their freezing and thawing data may be skewed. They make daily observations to report ice-on dates, when ice completely covers the lake, bay, or river and stays intact for the winter, and ice-off dates, when the ice completely disappears from the body of water for the year.

These observations—more than twenty-one thousand so far—can help to validate remote observations, including confirming the accuracy of space-based observations of ice freezing, Doyle says.

⍟ AGE RANGE All ages **▦ TIME FRAME** Winter **⊕ REGION** Canada **$ FEE** None

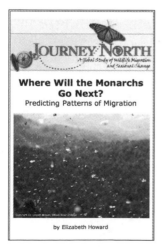

Where Will the Monarchs Go Next?
Predicting Patterns of Migration

by Elizabeth Howard

Journey North

www.learner.org/jnorth

Sponsored by Annenberg Learner, Journey North coordinates the observations students and citizen scientists throughout the northern hemisphere make about wildlife migration and seasonal changes. Participants share their observations of several migrating animals, including monarch butterflies, gray whales, and whooping cranes, as well as natural phenomena such as the

One of the many educational booklets available on the Journey North website.

first sighting of earthworms or the first sap run of maple syrup.

WEATHER AND THE SEASONS

"It's a great way to observe natural events and how the food web comes back to life every year," says founder and director Elizabeth Howard.

The program was originally targeted toward students. The website contains extensive information about the biology of the plants and animals studied in the project, as well as lesson plans for teachers. Students can also pose questions through the website to experts on each subject.

It's not limited to students, however. Howard estimates that about 40 percent of participants are families, nature centers, or other interested parties.

The project was designed to be simple for all ages. Citizen scientists can contribute sightings of each of these animals through the project's website or its mobile app, with no additional time commitment required. "It sets the tone as being really simple and welcoming," Howard says.

Even so, the information generated is carefully vetted and has been used in a number of scientific papers. And Journey North's wide network of observers (some eight hundred thousand people participate each year) provides value as well. "They are making observations that might otherwise go unnoticed," Howard says, such as the effects local storms or a warm spring might have on migrations.

The site generates maps in real time showing the seasonal progression of each of the phenomena the project follows. "There's a whole online community watching the same things," Howard says. "It's a great way to keep in touch with relatives or friends."

AGE RANGE All ages; particularly well-suited to young children **TIME FRAME** Year-round, particularly spring and fall **REGION** North America **FEE** None

---

Nature's Notebook   www.usanpn.org/nn/become-observer

Created by the USA National Phenology Network, Nature's Notebook collects information about plant and animal phenology—the timing of life cycle stages and their relationship to weather and climate. "Because of the tight relationships between climate conditions and plant and animal phenology, the timing of plant and animal events are changing with changing climate conditions," says Theresa Crimmins, USA-NPN partnerships and outreach coordinator.

Participants choose their own sites to visit regularly and make observations. They also choose the plants or animals to collect data about. Participants report which life cycle stages the organisms they are observing are in. For plants, participants record information about the same specimens each visit, while for animals, they record the presence or absence of each of the species on their site checklist.

It's an activity that is easily accessible to all ages. "We generally present Nature's Notebook as appropriate for ninth grade and up, but if kids and parents are doing it together, it is completely suitable for younger children," Crimmins says.

Participants report data through the Nature's Notebook website. A free account is necessary. Data is accessible by researchers, land managers, decision makers, and the public through the project's website.

The USA National Phenology Network is also home to an effort to crowd-source the digitization of historical phenology data. The North American Bird Phenology Program has six million Migration Observer Cards that contain information about bird migration arrival dates between the late nineteenth century and World War II. Volunteers are needed to scan or transcribe those cards into a computer database.

AGE RANGE All ages   **TIME FRAME** Year-round   **REGION** US   **FEE** None

Resources

ADULT BOOKS

Blame It on the Rain: How the Weather Has Changed History by Laura Lee (William Morrow, 2006).

Braving the Elements: The Stormy History of American Weather by David Laskin (Doubleday, 1996).

The Encyclopedia of Weather and Climate Change: A Complete Visual Guide by Julianne Fry, Hans-F. Graf, Richard Grotjhan, Marilyn Raphael, and Clive Saunders (University of California Press, 2010).

Weather Matters: An American Cultural History since 1900 by
Bernard Mergen (University Press of Kansas, 2008).
Windswept: The Story of Wind and Weather by Marq de
Villiers (Walker, 2007).

CHILDREN'S BOOKS

Experiments with Weather and Climate by John Bassett (Gareth Stevens Publishing,
2010). Ages 9 and up.
The Kids' Book of Weather Forecasting by Mark Breen and Kathleen Friestad (Ideals,
2008). Ages 8–11.
The Magic School Bus and the Climate Challenge by Joanna Cole and Bruce Degen
(Scholastic, 2010). Ages 7–10.
The Story of Snow: The Science of Winter's Wonder by Mark Cassino and Jon Nelson
(Chronicle, 2009). Ages 5–9.
*A Warmer World: From Polar Bears to Butterflies, How Climate Change Affects
Wildlife* by Carolyn Arnold (Charlesbridge, 2012). Ages 9–14.
Weather Projects for Young Scientists by Mary Kay Carson (Chicago Review, 2007).
Ages 9–12.

AFTERWORD

ABOUT THREE YEARS AGO, I BEGAN VOLUNTEERING AT THE PEGGY NOTE-baert Nature Museum in Chicago. When I started, I thought it might be a fun thing to do, but I didn't have terribly high expectations. I figured I'd enjoy the fantastic butterfly haven and talking to visitors about them, and that I'd feed the turtles every so often, and that once in a while I'd handle a snake (or, more likely, persuade one of the other volunteers to handle a snake for me).

At the time, I never realized the depth of what I would learn. I had no notion of just how much there was to understand about butterflies—the amazing variety of species, the fascinating stages of life that they go through, and even the incredible, often jewellike chrysalides that the caterpillars form before emerging in their final form. I didn't expect to care about the different types of turtles, or about the techniques of rearing endangered varieties in hopes of releasing them to the wild and rebuilding their population. And I certainly didn't think I'd ever become comfortable holding the snakes, let alone learn the differences both in biology and in personality between Coco (one of the museum's fox snakes), Connie (a corn snake), and Bianca (a bull snake).

Working on this book was a similar experience. I had a pretty good understanding of the mechanics of citizen science (which, coincidentally, I had first experienced at the Nature Museum), but I didn't have any sense of the passion that participants felt for it. I didn't realize that people kept with projects for decades, or that children who grew up with a project might eventually make a career out of the subject. It was a delightful discovery to make.

There is, it seems, citizen science research in just about any topic you might be interested in, and the number of projects is only likely to continue to grow in the foreseeable future. That's a wonderful opportunity for everyone—parents, children, grandparents, families, individuals, or groups. Learning about science and nature doesn't just teach you facts, and it's not about passing a test. It's about asking questions, even impossibly hard ones, and knowing that there is a way to eventually find an answer if you look at the question in the right way. It makes the world a much bigger and much more exciting place.

I hope you're inspired to take every opportunity that comes to see what's out there.

INDEX